PREFACE

The Juntura Basin is a subsidiary area within the Northern Great Basin of Western North America. It was the site of deposition of the Miocene and Pliocene rocks peculiar to it. Today this same area is made up of a series of small basins, hills, and uplands. A total relief of over two thousand feet obscures its former features. The present topography is for the most part a result of folding and faulting of the rocks accumulated during the Miocene and Pliocene. Some features are much older and represent irregularities in the original surface upon which the rocks we have studied accumulated. These old ridges are found in the eastern part of the basin. They indicate along with other information that the Juntura Basin was probably never a simple smooth-floored area. Accompanying the evolution of the topography, the climate, flora, and fauna have changed. The warm moist climate of the later Miocene has gradually given way to the harsh climate seen there today. Slow-moving streams and occasional lakes were occupied by beavers, storks, and flamingos. On the adjacent land and marshes lived camels, horses, rhinoceroses, and a host of other animals whose relatives are not seen there today. These spectacular changes catch the imagination of both scientist and layman.

Our studies thus far have been confined to the northerly part of the Juntura Basin. Its southerly limits have not yet been precisely determined. Little was known of the physical and organic history of this area when we first started work there. Because of this, much of the work presented in this report is descriptive. Several recent publications concerning this area are interpretive (Chaney, 1959; Shotwell, 1958, 1960; and Brodkorb, 1961)[1] or are concerned with problems of a broader nature geographically. Chapter 1 of this report interprets the paleoecology of the late Tertiary, primarily that of the mammals.

Chapter 2 is a preliminary report on the geology of the basin confined largely to the nature and distribution of the rocks. Studies on the origins of the rocks are in progress. The master's theses of three students, Bowen, Gray, and Gregory are combined to provide these data. I have attempted to retain as much of their original wording and ideas as possible and still form a single consistent work.

We do not consider our work in this area to be completed. The present report includes a great deal of information and provides solutions to some of the problems we saw initially. There are, however, significant problems yet to be solved both geologically and paleontologically. The record of earth history preserved in this relatively small area is by no means completely understood. This report is in many respects an indication of progress in a long-term effort.

ACKNOWLEDGMENTS

I wish to express my appreciation for a great deal of hard physical work, enduring patience, confidence, and helpful criticism to my wife, students, colleagues, friends, and to granting institutions. Students, undergraduate and graduate, who have aided in the field work and preparation are: Huntley Alvey, Morton Berenson, Ronald Bigelow, Richard Bowen, Douglas Burns, John Ellis, Wilfred Gray, Dolores Gregory, John Patrick, Ray Reeder, Dale Russell, Don Russell, and Alex Veetch.

The American Philosophical Society provided funds with which this program was initiated (Grants Nos. 1700, 1834). The University of Oregon Research Committee helped with needed equipment items and funds. The National Science Foundation provided funds which made it possible to continue work on a larger and more productive scale (1955–1957).

The California Institute of Technology (CIT) and the University of California at Berkeley (UC) have allowed me the free use of their collections and field data. I am grateful to those in charge of the collections and to those who made them.

Little of this work would have been possible without the aid and consideration received from the people living in the Juntura Basin. Mr. and Mrs. F. C. Gurney of Juntura were our first acquaintances there. Their position of respect in the community made it possible to meet, under ideal circumstances, many other people of Juntura and the surrounding ranchland. A paleontologist is always dependent on the assent of landowners to work on their property. I am, therefore, grateful to Drex Williams, Raymond Joyce, Glen Sitz, the Butlers, Allens, and many more for this permission and for many other considerations. I am indebted to a number of other local residents and to them extend my thanks. Mr. Al Brown of Burns, Oregon, has aided us in a number of ways. These people have become friends whose friendship I shall value long after the Juntura Basin has lost, if ever, its importance to me as an area for the study of Earth History.

Mildred R. Detling prepared the drawings in figures 37–119. Robert Stevens drew the maps.

J. A. S.

[1] See list of references at end of chapter 1.

THE JUNTURA BASIN: STUDIES IN EARTH HISTORY AND PALEOECOLOGY

J. Arnold Shotwell

With Contributions by

R. G. Bowen, W. L. Gray, D. C. Gregory, D. E. Russell,
and Dwight Taylor

CONTENTS

1. PLIOCENE MAMMALIAN COMMUNITIES OF THE JUNTURA BASIN

J. A. SHOTWELL

Museum of Natural History, University of Oregon

INTRODUCTION

The mammalian fauna of the Great Basin of Western North America has changed dramatically since the late Miocene. Of thirty-one phyletic lines present in the Northern Great Basin in the Miocene, only a half dozen are still represented. Many new mammals appear as migrants from other continents and other regions of North America. Changes in the composition of the flora are equally impressive. The woodland and forests of the Miocene are depleted until in the mid-Pliocene they are represented only along stream drainages and at higher elevations, with savanna and open grassland occupying much of the area once forested. In the later Pliocene this trend is temporarily reversed (Chaney, 1959; Axelrod, 1950, 1960). These changes of vegetation reflect changes of climate and to a lesser degree topography. New habitats for mammals are thus created as old ones are reduced or lost. There is a significant evolution of mammals in this late Tertiary segment of time; however, the changes in the mammalian fauna from late Miocene to late Pliocene are apparently due primarily to environmental changes and, to a lesser extent, to evolution of residents. This sequence presents an excellent opportunity to study the roles of environment and evolution in faunal change.

The history of groups of mammals related environmentally has seldom been studied or even recognized. The community is such a group and is useful in the study of the history of mammals. Changes in the organization within communities can be studied as well as their beginnings, extinction, and sometimes their distribution. Paleoecology is thus concerned with the evolution of communities.

The study of communities of mammals, their composition, structure, and history, must be quantitative and sequential in order to be significant. Shotwell (1955) described a technique to recognize the members of a community in samples from local concentrations of fossil mammals in the sediments. This approach was applied to a series of samples of the same age and from the same geographic area to determine its sensitivity to the contemporary diversity of communities and their interrelations (Shotwell, 1958a). A sequence of such studies in geologic time can provide the basic data for the study of the evolution of communities. Sequential studies must be carried out within a biologic province and with adequate geologic controls. Biogeographic differences can be avoided by restricting work to a single biogeographic province. When the samples studied in a sequence are in a superpositional relationship, it is not necessary to determine the time order of samples in the sequence from the samples themselves. When these requirements can be met, many difficulties of analysis can be avoided. It is seldom possible to have more than two or three successive steps in a sequence available locally. However, if one of these is repeated in a nearby basin in the same biologic province, plus additional steps, a considerable sequence can be made available. Geologic control between such portions of a composite sequence is highly desirable. The Northern Great Basin has been the region of primary interest in these studies because of the occurrence of sequences which meet the above requirements. Available information concerning late Tertiary vegetational change makes this province even more desirable for studies in the paleoecology of mammals.

The well-established terms of the vertebrate chronology will be used in this paper. A list of those of the late Tertiary and their approximate equivalents follows: Barstovian (late Miocene), Clarendonian (early Pliocene), Hemphillian (mid-Pliocene), Blancan (late Pliocene).

Our early efforts to find a suitable sequence were disappointing. Frequently we found sequences but they often lacked key units or these key units were not productive. We finally were successful in finding a Pliocene sequence near the town of Juntura in Malheur County, Oregon. Hemphillian beds were locally superimposed on those of Clarendonian age. These, in turn, overlay Barstovian beds. This local basin which we have subsequently named the Juntura Basin is the key to our sequential studies. The Barstovian beds in the Juntura Basin have yielded only plant fossils so that our mammalian sequence is only the minimum two successive steps; Clarendonian and Hemphillian. Subsequent to our work in the Juntura Basin we have established the Barstovian and Blancan steps in what we hope will be controlled sequences related to the Juntura Basin. Much of the excavation and geologic studies on these additional steps in the sequence have been completed and will be described at a later date.

The Juntura Basin is on the very northern border of the Great Basin. The Malheur-Harney county line passes through it north-south and U. S. Highway 20 passes through it east-west. (See map 1.)

METHODS

The fossils collected from the Juntura Basin include plants, fresh-water mollusks, fish, reptiles, amphibians, birds, and mammals. Each of these groups other than the mammals was sent to a specialist to be studied.

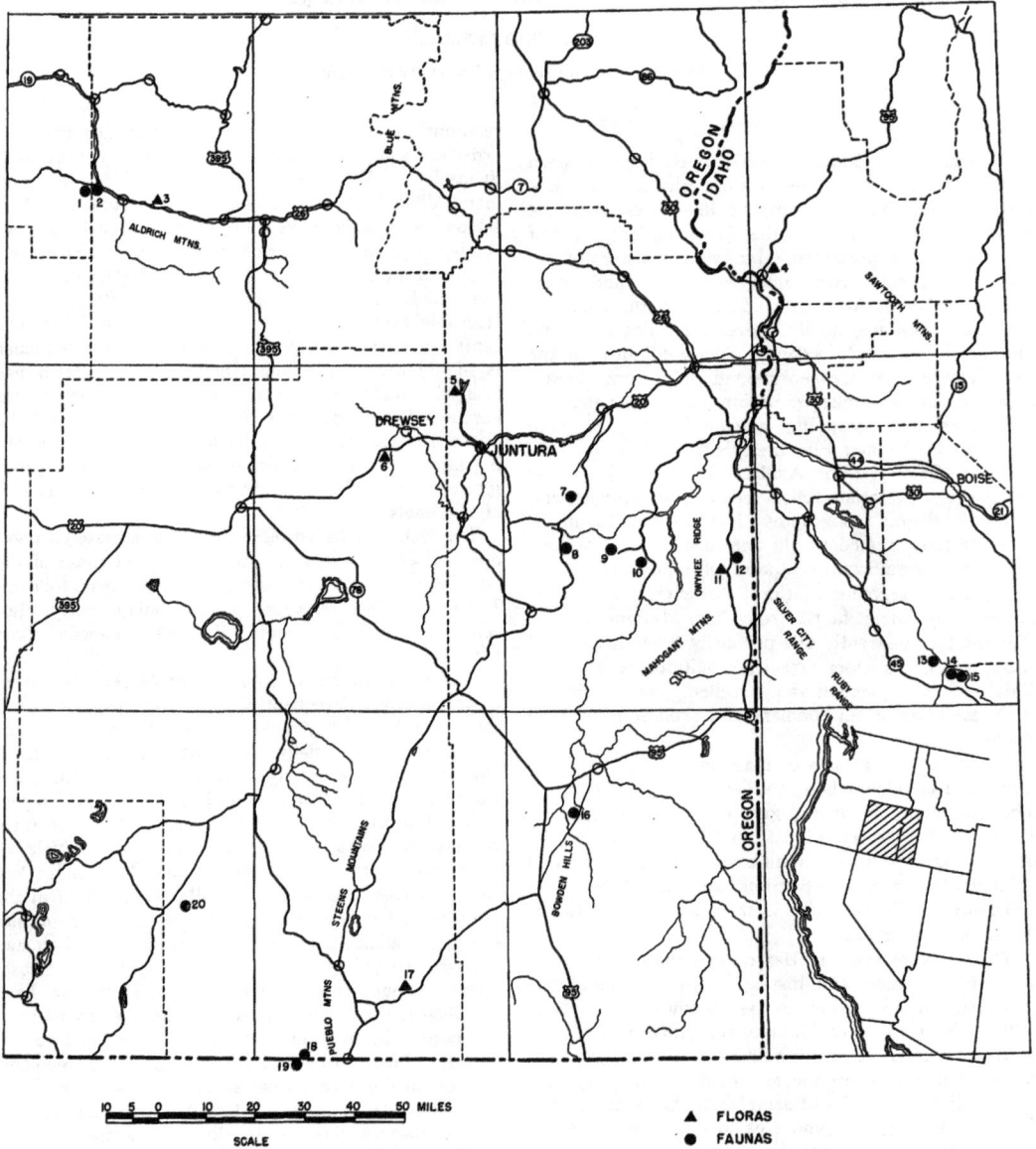

MAP 1. Localities in areas adjacent to the Juntura Basin.

1. Rattlesnake
2. Mascall
3. Mascall
4. Weiser
5. Beulah
6. Stinkingwater
7. Skull Springs
8. Skull Springs
9. Juniper Creek Canyon
10. Quartz Basin

11. Sucker Creek
12. Sucker Creek
13. Wildhorse Butte
14. Castle Butte
15. Jackass Butte
16. Rome
17. Trout Creek
18. Thousand Creek
19. Virgin Valley
20. Beatty Butte

Reports on some of these appear in this volume. Others will appear at a later date and some have already appeared. Study of the geology of the basin was carried on concurrently with the excavations.

The mammalian samples were analyzed as described by Shotwell (1955, 1958a). The primary goal of this approach is to provide an objective basis by which the mammals which lived together in a community may be segregated from those present in the sample owing to other factors. This approach is designed to delineate the communities represented and indicate the relative abundance of members of the major community at each locality. The adaptive morphology of each of the members of the major community suggests the nature of the habitat requirements of the community. Plant materials are seldom found in the quarries with the mammals, but study of the various plant associations recognized from nearby localities indicates the nature of the vegetation of the habitat indicated by the adaptive morphology of the mammals. Conditions of climate, topography, and the presence and nature of water bodies are determined from those materials to which can be applied a strict uniformitarian transfer of present requirements of living morphological equivalents. Since mammals change throughout the late Tertiary, they cannot be used in this sense. The character of the sediments provides additional important evidence and will yield more if our early attempts at this continue to be successful. Mammals in the sample which are shown to be from communities other than the major one indicate something of the nature of adjacent areas and the fact that such communities were also present but do not provide usable quantitative information concerning their own community. Samples must be studied in which these mammals are members of the major community to determine their associates and relative abundance. Examples of this are described in Shotwell (1958a).

Wilson (1960: 9) has discounted the method applied here considering it to be too much work and concludes (p. 10) that "it remains to be demonstrated how much more accurate the method proposed by Shotwell is than is one using percentage calculations derived from numbers of specimens, with perhaps some subjective adjustment for other considerations." He considers the theoretical basis for the method to be sound but then suggests a technique which has no theoretical basis. If such a method, using percentage of number of specimens, is applied to the same samples, the results are quite different from that obtained by my approach in the ordering of species. No indication is given as to how this will indicate which mammals were associated in life and which are foreign to the community. It does not provide any basis for recognizing low-abundance forms in the community as opposed to fragmentary remains of a high-abundance form from a distant community. In other words a percentage ordering of the species does not provide any information

about a community. This is the primary purpose of my method.

The method I have employed requires about one hour of time to make the calculations. The excavations, preparation, identification, and counting of specimens requires a great deal of time, but these are procedures which are necessary in any quantitative study.

Wilson has suggested that short cuts can be made by not using all the available sample and by not allowing for the bias in data due to the differences between taxonomic groups of mammals in the number of skeletal elements that can be contributed to a deposit. If such short cuts are used, the method is then no longer quantitative and certainly not accurate. My method allows for difficulties in assignment of specimens but this cannot be extended to the point where no post-cranial skeletal elements are used as Wilson has done (Wilson, 1960).

Wilson states (1960: 9) that a "check" he has made of my method "proved disappointing." The only test of a method such as the one I propose is one which determines how accurately the results describe the original community. Nothing in Wilson's discussion approaches such a possibility. Since we are dealing with paleo-communities, direct testing of the method may be difficult but it can be tested on the consistency of its results and its theoretical basis. I have continually cross-checked results and have applied the method where local diversity was apparently present to see whether the results reflected this (Shotwell, 1958a). Wilson apparently has no argument with the theoretical basis. His disappointment came when he studied bird collections from Fossil Lake and McKittrick. He assumes that birds accumulate in the same way as mammals. Paleontologists have frequently commented on the rarity of volant forms in deposits and considered that this does not reflect the abundance of these forms at any one time but is a result of their means of locomotion which for some reason keeps them from interment. Be this as it may, the collections from Fossil Lake were made beginning in the late 1800's and are a composite from a number of localities. No effort was made to accumulate controlled samples. We do not know under what conditions the McKittrick collections were made. These samples, however, do not satisfy the primary requirement of a quantitative sample that it be all the materials from a given volume of sediment in a single lithology and that it be non-selective. McKittrick samples are undoubtedly highly selective as was pointed out in the case of mammals from Rancho La Brea which represents a similar type of accumulation (Shotwell, 1955).

We can see no basis in Wilson's remarks for abandonment or modification of the method employed here or its replacement by an easier one which has no theoretical basis.

Local concentrations of fossil mammal bones in place in the sediments have been difficult to find in the

Northern Great Basin. This has not only been a problem to us but to all others who have worked in this region. The common occurrence of considerable amounts of fragmental material on the slopes and in gullies announces the presence of bone but frustrates the worker when he attempts to discover the source. Slopes are often steep, frequently 27°, and are covered with a thin mantle of weathered sediment. We have finally settled on a technique to discover the source and occurrence of concentrations of bone in the sediment. It requires a good deal of time and patience mixed with effort, but is eventually successful. This technique often involves two or three weeks of work after we have discovered surface bone on the slopes in an area.

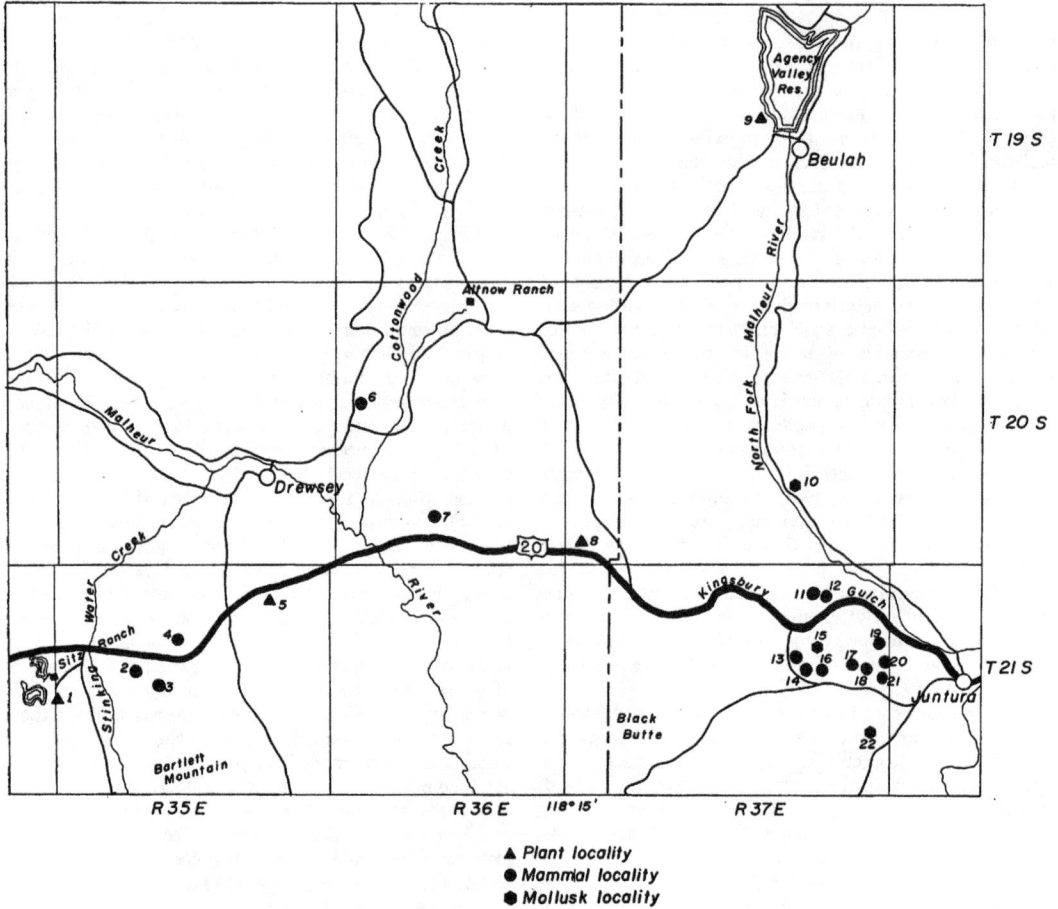

▲ Plant locality
● Mammal locality
⬣ Mollusk locality

MAP 2. Major localities in the Juntura Basin.

1. UC P4120
2. UO 2356)
3. UO 2355)==CIT 107
4. UO 2239)
5. UO 2367
6. UO 2347
7. UO 2361
8. UO Plant locality
9. UO 2330
10. UO 2363=USGS 19116
11. UO 2326

12. UO 2332, Q2
13. UO 2337
14. UO 2336
15. UO 2371=USGS 19117
16. UO 2340
17. UO 2448, Q3
18. UO 2352
19. UO 2335
20. UO 2360
21. UO 2344
22. USGS 21173

Very few of the quarries we have established in the Juntura Basin or in subsequent areas of interest have been found by more direct means, such as protruding bone or accumulations of bone at that point. The steep slopes with an apparently moving mantle of weathered sediment do not often allow such accumulations. Protruding bone is sheared off and carried along in the mantle for some distance before it is exposed on the surface. Our technique is an attempt to follow this bone back up through the mantle to its point of origin. This is accomplished by carefully scraping away the mantle in a strip about four feet wide proceeding up the slope. Bones are usualy chosen which indicate by lack of bleaching or surface weathering that they have not been exposed long. Complete bones are of little aid in this technique since they leave no fragments to mark their movements from their point of origin. They can, however, indicate by their completeness that they have moved only a short distance. Exploring around them under the mantle may expose others. Large or long skeletal elements are most useful since they leave numerous fragments as a trail and show up more readily in the trowelling. A number of teeth of the same species, which by their size and wear suggest they are from the same individual, may serve the same purpose. If the site is producing many small mammals, these can be traced with care by the same process. When bone is finaly found in place, it is necessary to determine whether it is a single occurrence or whether a concentration is present. A test pit must then be cut to see whether other bones are present. This can be done rather hurriedly since the aim is to determine the presence or absence of a concentration and not the beginning of its study. Since concentrations are not usually homogeneous in their occurrence, this test pit should be extended laterally within the lithology of the original find and into the slope with some persistence until it is clear that the necessary concentration is or is not present. When the presence of very small vertebrates is suspected, tailings should be dry or wet screened, depending on the nature of the sediment, to avoid missing a concentration. Often in our experience we have tracked down larger bones to their origin and found that the most abundant specimens present in the concentration were very small and that larger material was rare. These concentrations could have been overlooked without screening.

Once we are reasonably sure that we have located a promising site we organize our approach largely according to the nature of the sediment in which the concentration occurs. This primarily is indicated by the type of screening to be carried out, wet or dry. Wet screening serves no purpose in loose, coarse-bedded sand since it is not necessary to break down the matrix. Dry screening is slow since the concentrate must be sorted in the quarry, whereas with wet screening the concentrate is dried and hauled back to the museum for sorting. Many more problems of logistics are involved in the handling of wet screens although it usualy goes faster. The productive matrix must be hauled to a suitable site where water is available, sometimes a real problem in semi-arid country. A portion of the crew is assigned to this duty. Access to the quarry itself must be better in order to move the many tons of sediments. We often use a high-line across a gully or cut switch-backs up the hill for a jeep. Our screens both dry and wet are made of screening 14 × 18 squares to the inch, folded as a box and attached to a rectangular loop of ¼-inch rod by means of hog rings. This alows a free flow of water in and out of the screen. The screens are made up as they are needed, usually about eighty in number for wet screening. At the end of the season the screen is removed from the rods and discarded. The rods are then tied in bundles and are thus simple to load for hauling to and from the museum.

When the vertical limits of the productive layer are determined, the overburden can be attacked. On steep slopes this is often a major problem if we are to expose and remove a large volume of productive sediment. We generally cut the overburden off down to the proximity of the productive layer by means of a jack hammer using bits (sizes 1–6 inches) appropriate for the hardness of the rock. As the tailings pile up they are removed by a scoop attached to a winch on one of the vehicles. This results in a quick, clean job, and leaves the productive level ready for more demanding excavation. This step may be carried out a number of times and usually is limited by the lateral extent of the concentration or the safe height that can be maintained on a back wall. We limit ourselves to 20 feet but often need only 15 feet of back wall. Excavation for quantitative samples cannot be carried out with a small crew or no mechanical equipment unless you have free help and a very long season. Savings in wages over hand work soon pay for the necessary equipment. The gain in productive field time cannot be evaluated. We have come to these conclusions after trying both ways.

Dry screening is carried on in the pit using three screen meshes, four, eight, and fourteen squares to the inch. This hurries a tedious job by allowing the sorting of things of approximately equal size.

Wet screening often presents problems which must be solved as they appear. Most matrices must be alternately soaked in water and dried in the sun until the concentrate is reduced to a point that it can be transported economically. Each matrix requires somewhat different procedures. Hibbard (1949) has successfully used this technique and suggests means for handling such problems.

A primary concern, but one easily overlooked, in the work of removing bone from a quarry is the nature of the sediment in which the bone has accumulated. Some characteristics are obvious, that is, coarseness and presence or absence of bedding, etc. Many other very important aspects of these sediments must be determined by careful examination. Such characteristics as nature of sorting, heavy mineral content, data of paléocurrent direction are of utmost importance. We are presently studying a series of quarries we have excavated to determine whether there is a correlation of conditions of accumulation of the sediment and the nature of the community found there. Preliminary studies show that such a correlation is in fact the case. These studies combined with stratigraphic and sedimentary petrographic studies of the units in which we find our concentrations will greatly increase our knowledge of the environment of deposition and the nature of the accumulation of concentrations of bone in a deposit.

ORIGIN OF THE BLACK BUTTE FAUNA

The Clarendonian Black Butte fauna of the Juntura formation is described in detail in chapter 4. We shall be concerned here with its origin and the communities represented. As noted previously, the underlying Barstovian sediments have not produced any mammals but do contain abundant plant materials. We must look to closely adjacent areas for Barstovian faunas. Review of these faunas indicates what portion of the Black Butte fauna is derived from Northern Great Basin residents of the Barstovian and what portion represents probable new types to the region at this time. The Skull Springs fauna (Gazin, 1932) and the Quartz Basin fauna at Ferguson Springs (both Barstovian) are located to the southeast of the Juntura Basin at a distance of about twenty-five and forty airline miles respectively. These faunas represent a number of environmental situations. Mammals in the Black Butte fauna whose probable origin is in the preceding resident Barstovian fauna can thus be easily recognized. Numerous other Barstovian faunas at greater distances, but yet within the Northern Great Basin, are of additional aid and represent still other environments. Twenty-one of the twenty-eight Black Butte genera have their probable ancestors in the Skull Springs or Quartz Basin faunas. Four additional genera (mustelid carnivores) may have their origins in those few mustelids known from the Barstovian faunas of the Northern Great Basin. The state of our knowledge of this group makes it difficult to determine whether these mammals have diversified from mustelids represented in Northern Great Basin faunas or are from other regions. Similarly the origin of *Peromyscus* is in some doubt although a very possible ancestor is present in the Quartz Basin fauna. *Tardontia* is a rare form known in the Barstovian only from the Central Great Basin. It very probably was a resident in the Northern Great Basin in the Barstovian.

Only the shovel-tusk mastodon, *Platybelodon*, is an obvious migrant into the region in the Clarendonian. The Black Butte fauna then has its origin primarily in the earlier residents of the region. It differs from the previous fauna in the lack of browsing horses,

primitive dogs, and bear dogs, and a significant reduction of artiodactyl groups, primarily browsers. A number of the phyletic lines of the Barstovian which continue into the Black Butte fauna are more diversified in this later fauna, viz. aplodontid rodents, mylagaulid rodents, beavers, camels, and possibly mustelids. Migrants are an insignificant portion of the Black Butte fauna.

COMMUNITIES OF THE BLACK BUTTE FAUNA

Communities are delineated here by quantitative studies. The nature of the habitat occupied is indicated by the functional morphology of all the mammals present as well as other vertebrates present in the sample. Thus amphibious mammals demand aquatic habitats or at least habitats where water bodies are an important part, browsing mammals require broadleafed vegetation, mammals with high-crowned teeth and appendages well adapted to high-speed running indicate grazing conditions. There can, however, be more than one community type in such broad categories of habi-

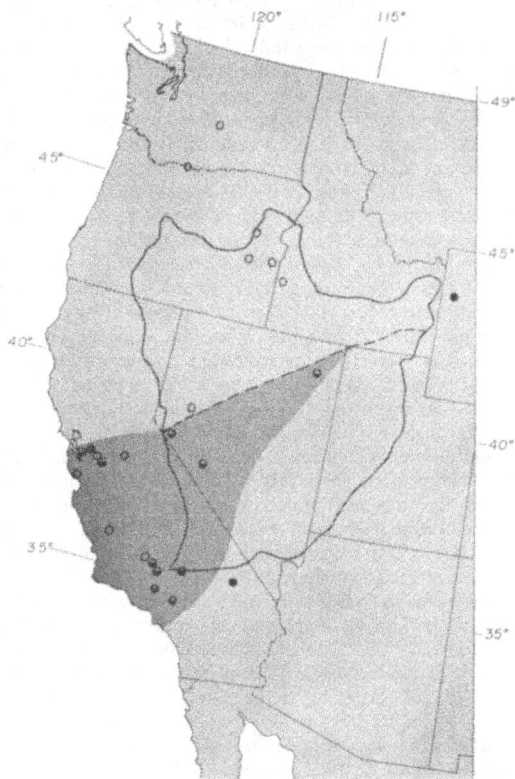

MAP 3. Distribution of *Hipparion* and *Pliohippus* in the Clarendonian of Western North America. \ \ \ \ *Hipparion*, / / / / *Pliohippus*.

tat. For instance, grazing communities may be savanna or open grassland, etc. The recognition of these differences and the nature of the habitats reflected often requires the use of additional information. In the Pliocene of the Northern Great Basin at least two grazing communities are recognized. Their primary differences as seen in the samples are in the horses present in each and to a lesser extent some of the artiodactyls. They apparently have a number of species in common as might be expected. These communities may be characterized as the *Hipparion* (three-toed horse) and *Pliohippus* (single-toed horse) communities for the sake of discussion here. The distribution of these horses in the Clarendonian (early Pliocene) and Hemphillian (mid-Pliocene) in western North America points out some interesting facts which must be accounted for in order to understand what differences may exist in the habitats preferred by the communities of which these horses were prominent members. Maps 3 and 4 are distribution maps of *Hipparion* and *Pliohip-*

pus in western North America in the Pliocene. Several important points can be made from these maps.

1. There are large areas in which only one or the other of the horses occurs, that is, areas of mutual exclusion.
2. There is an intervening area in which both occur often at the same localities.
3. The southern limit of the distribution of *Hipparion* retreats to the north and west in the interval of time considered.
4. The northern limit of *Pliohippus* extends northward during the Pliocene.
5. The area of mutual occurrence of the two horses moves north and west in the Northern Great Basin.

Some of these biogeographic changes were previously noted (Shotwell, 1961). The chief differences in the distributions shown here and those described earlier are in the recognition of areas in which *Hipparion* apparently did not occur. Subsequent excavations have demanded such a conclusion. This additional information, however, does not affect the conclusions reached earlier but provides additional evidence useful in the determination of the habitat preferences of the communities to which these horses belonged.

The points made by the maps provide strong evidence that at least two communities are involved and possibly more. As noted previously, the extension of *Pliohippus* to the north parallels the apparent movement of open grassland into the Northern Great Basin (Shotwell, 1961). Paleobotanical evidence for the appearance of open grassland is indirect. However, the retreat of *Hipparion* to the north and west correlates closely with the well-documented retreat of woodland and forest species in the Great Basin flora. This additional agreement strengthens my earlier contention that *Hipparion* horses preferred grazing situations associated with woodland or, in other words, savanna-woodland. I have called this simply savanna.

The areas shown on the maps in which both species occur may represent merely close proximity of these habitats and thus the frequent mixture of communities in samples. This more than likely is the case in the Northern Great Basin. However, it may be that a more complex savanna community was present in California such as a lowland savanna. Several mammals associated with savanna or open grassland of the Great Basin do not occur in the California area suggesting basic community differences.

The above discussion emphasizes that the historical biogeography and vegetational history in addition to the functional morphology of members of a community must be examined in order to understand the differences in communities of similar habitat requirements.

Two of the many Clarendonian sites established in the Juntura Basin provide quantitative information suitable for community analysis. A third has provided

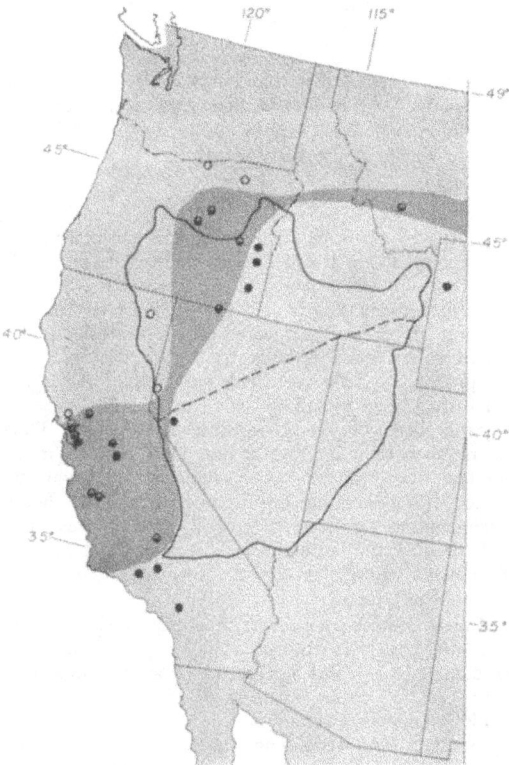

MAP 4. Distribution of *Hipparion* and *Pliohippus* in the Hemphillian of Western North America. \\\\ *Hipparion*, //// *Pliohippus*.

a small sample which apparently represents the same community as one of the other two.

THE SAVANNA COMMUNITY

Quarry 3 (UO loc. 2448) of the Black Butte Fauna was established in a lens of cross-bedded sand. The data of this sample are seen in table 1. These data are the basis for the faunal analysis diagram (fig. 1). All of the mammals in this sample have relatively high ratios of corrected number of specimens to number of individuals. The lowest is approximately the average of other samples. This indicates that the sample is probably one in which mammals from foreign communities are not present. The mammals present belong to a single community. They include *Procamelus* (camel), *Megatylopus* (large camel), *Hipparion* (three-toed horse), *Mammut* (mastodon), and *Aphelops* (rhinoceros). This community, represented in the sample from quarry 3, is undoubtedly the savanna community of the Clarendonian in the Juntura Basin. Our sample apparently only contains the most abundant and larger forms, since carnivores and rodents are lacking. A common carnivore to be expected in such a community and present in those previously studied is the hyaenoid dog *Osteoborus*. This dog is present at other local sites in the Juntura formation where inadequate samples for analysis were collected. It should probably be included in the membership of this community. A quarry producing a small sample (Q2) apparently represents the same community. It indicates the presence of still another carnivore, a large hyaenoid dog, *Aleurodon*. Other mammals known from the various localities making up the Black Butte fauna were probably members of the savanna community, but there is no basis in our available data for including them at this time. The frequency of the occurrence

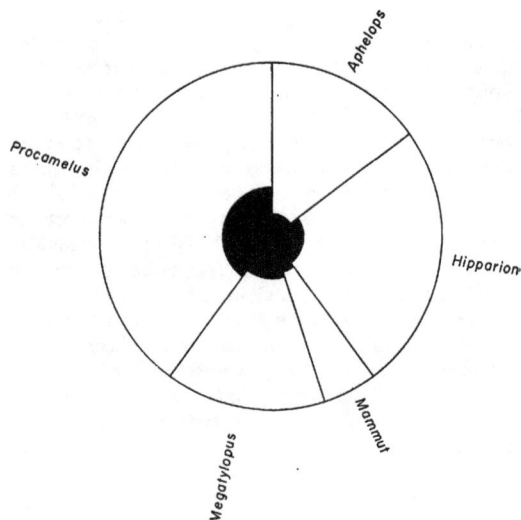

FIG. 1. Faunal analysis diagram from sample at quarry 3, UO 2448.

of the members of this community in float collections and small associated collections indicates that this community may have been a predominant one in the Juntura Basin in the Clarendonian.

POND-BANK COMMUNITY

Quarry 11 (UO loc. 2337) was established near the top of a small knoll in a fine-grained tuffaceous sandstone low in the upper Juntura formation. Matrix from this quarry was hauled to a suitable site on the Malheur River and wet screened. The residue was sacked and returned to the laboratory for sorting. This site produced 13,200 vertebrate specimens. As is the usual practice in vertebrate paleontology, disassociated elements are considered separate specimens. Thus the number of individuals represented here is much less than the total number of elements and fragments of elements making up the figure cited above. These specimens represent several groups of vertebrates: fish, birds, turtles, and mammals. The relative abundance of elements of these groups is shown in a bulk faunal analysis in figure 2.

Mammals make up a rather small part of this sample. The faunal analysis diagram (fig. 3) constructed from the data of table 1 shows *Eucastor* (beaver) and *Hypolagus* (rabbit) to be members of the proximal community, the major community adjacent to the site of deposition. The remaining mammals: *Hesperosorex* (shrew), *Scapanus* (mole), *Peromyscus* (white-footed mouse), *Cupidinomys* (heteromyid mouse), *Citellus* (ground squirrel), *Vulpes* (fox), mastodont, *Ustatochoerus* (oreodont), and *Prosthennops* (peccary), ap-

TABLE 1

QUARRY SAMPLE DATA BLACK BUTTE LOCAL FAUNA

	Q11 UO 2337			Q3 UO 2448			Q2		
	Number of specimens	Number of individuals	Corrected no. sp./ind.	Number of specimens	Number of individuals	Corrected no. sp./ind.	Number of specimens	Number of individuals	Corrected no. sp./inc.
Hesperosorex	1	1	5.9						
Scapanus	6	2	5.9						
Hypolagus	16	1	37.2						
Citellus	3	1	2.4						
Cupidinomys	1	1	2.4						
Hystricops							1	1	2.4
Eucastor	62	4	37.8						
Peromyscus	2	1	4.8						
Aelurodon							2	1	1.96
Vulpes	3	1	2.3						
Mammut	1	1	6.7	11	1	10.7			
Hipparion				44	5	8.5	18	1	17.5
Aphelops				25	3	6.3	2	1	1.5
Prosthennops	1	1	6.7						
Ustatochoerus	1	1	6.7						
Procamelus				108	8	13.6	34	2	17.3
Megatylopus				32	3	11.2			
Totals	97	15		220	20		57	6	

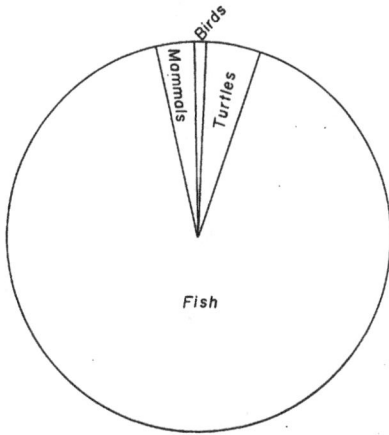

FIG. 2. Bulk fauna diagram from sample at quarry 11, UO 2337.

pear in the faunal analysis diagram as members of communities more distant from the site of deposition. Some mammals listed here as members of other than the proximal community may be members of this community owing to the effects of the size of the sample. Shotwell (1958a) has shown that this is likely to be the case in small samples. The small rodents and insectivores are probably the only forms affected in this instance.

The results of the study of the sample from loc. UO 2337 are strikingly like those obtained in the study of a younger (Hemphillian) sample from Krebs ranch (no. 1) along the Columbia River (Shotwell, 1958a: 277). All of the mammals seen in the Krebs Ranch sample or their immediate ancestors are found at UO 2337. The proximal community at Krebs Ranch includes *Hypolagus* and *Dipoides*. *Dipoides* is a beaver descendant from *Eucastor*. The samples are of similar size and bulk composition. The rabbit (*Hypolagus*) is not as abundant at the Black Butte site as in the later (Hemphillian) site.

The fish from loc. 2337 were examined by Dr. Robert R. Miller. They include catfish, squawfish, and sucker. These probably occupied a lake or a large river (personal communication to Russell). A more complete study of the fish by Miller and students now is in progress.

The birds from loc. 2337 as described by Brodkorb (1961) include cormorant, teal, and duck. Stork, flamingo, and goose are known from nearby localities.

The presence of the beaver, *Eucastor,* and the aquatic birds along with terrestrial forms indicates a pond-bank or slow-moving stream, large river habitat for the proximal community represented. The nature of the fish present indicates a very large pond or possibly a quiet stretch in a large river. The abundance of small

turtles emphasizes this. The Krebs Ranch sample referred to above was interpreted as occupying a similar situation.

Our quantitative samples have provided a basis for assignment of only about half the composite faunal list of the Black Butte fauna. Only one of the mammals not assigned to a community is represented by more than three specimens. Thus the bulk of the thousands of fossil mammal specimens collected is assigned. The one relatively common form unassigned is *Mylagaulus.* This rodent is also present in the Hemphillian fauna from the Drewsey formation. It is present there in a quantitative sample. Interpretation of this sample indicates that *Mylagaulus* is a member of a savanna or open grassland border community. Half of the unassigned mammals are carnivores. Their common occurrence in more than one type of community and their low relative abundance make their assignment dubious without very large samples. Small canids are almost always present in the samples of pond-bank communities. The possible occurrence of the fox, *Vulpes,* in the pond-bank community of the Black Butte fauna is then to be expected. It appears in our sample as a member of a distant community. *Osteoborus,* a hyaenoid dog, is present as a member of the grassland and savanna communities in large samples previously studied (Shotwell, 1958a). It is assigned to such communities here from these earlier studies. The other carnivores include a large cougar-sized cat, badger, marten, wolverine, and otter. No data from our samples at Black Butte or from former studies suggest assignment of these mammals to a specific com-

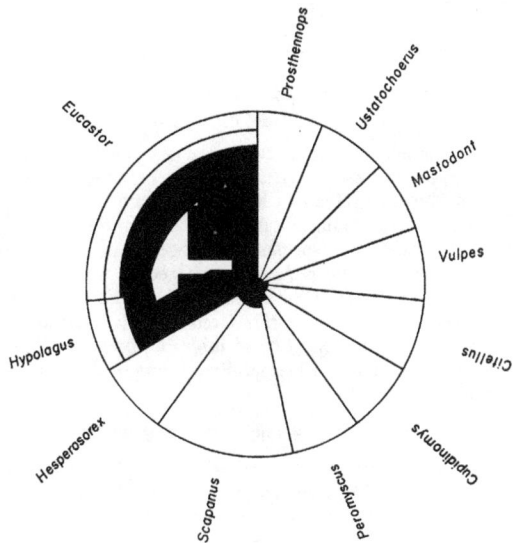

FIG. 3. Faunal analysis diagram from sample at quarry 11 UO 2337.

munity. *Prosthennops*, peccary, is rare in the Black Butte fauna. Previously it was found as a member of a number of communities ranging from pond-bank to savanna. The rodents *Tardontia, Epigaulus*, and *Hystricops* cannot be assigned at this time. The shovel-tusked mastodon, *Platybelodon*, occurred as isolated specimens in each instance. Its morphology suggests that it may be an occupant of marshy areas; however, we have no quantitative data by which we may assign it to a community. The two remaining genera, *Ustato-choerus* (oreodont) and the tapir, may represent woodland communities otherwise not represented.

A brief review of the unassigned material suggests that possibly one major community (woodland) and a minor community (border) were also present in addition to those represented in quantitative samples. The lack of open grassland ungulates (*Pliohippus* and antelope), which we would expect if open grassland were present, demands the conclusion that such a community was not yet developed. This question has been dealt with in detail by Shotwell (1961).

COMMUNITY ASSIGNMENTS

BLACK BUTTE FAUNA

Assignment is based on samples from the Juntura Basin except where indicated.

Savanna	Pond-bank	Woodland	Border	Unassigned
Procamelus	*Eucastor*	?*Ustatochoerus*	*Mylagaulus*	*Platybelodon*
Megatylopus	*Hypolagus*	?*Tapir*	?*Epigaulus*	*Hystricops*
Hipparion	*Hesperosorex*			*Tardontia*
Mammut	*Scapanus*			*Sthenictis*
Aphelops	*Cupidinomys*			*Pseudaelurus*
Osteoborus	*Citellus*			*Eomelivora*
Aleurodon	?*Vulpes*			*Pliotaxidea*
?*Prosthennops*	?*Prosthennops*			

*—Other basins. ?—questionable assignment from available evidence.

ORIGIN OF THE FAUNA OF THE DREWSEY FORMATION

The mammals of the Drewsey are described in a later chapter of this report (chapter 5). Of the nineteen genera known, twelve can be derived from the preceding Black Butte fauna. Another, *Liodontia*, has its apparent ancestry in species of that genus known from Skull Springs. The remaining six genera are new to the Northern Great Basin at this time. The faunas of the Drewsey formation differ from the Black Butte in the much greater proportion of migrants present. This is generally true in the Hemphillian fauna of the Northern Great Basin.

COMMUNITIES OF THE DREWSEY

We have only one quantitative sample from the faunas of the Drewsey formation. Fortunately our studies of Hemphillian communities along the Columbia River, to the north, provide a basis for assigning the materials in our collection with some confidence (Shotwell, 1955, 1958a). The study of community relationships and diversity referred to was carried out, in part, to supply the background necessary to interpret the results of studies in the Juntura Basin. Although these studies were concerned with an area north of the Great Basin, some of the same species and many of the same genera of mammals were present in both areas in the Hemphillian. Since only one sample large enough for a faunal analysis was obtained in the Drewsey formation, it is necessary to rely on the results of the studies on the Columbia River where possible in analyzing the Hemphillian communities which followed those of the Clarendonian (Black Butte fauna). Small Hemphillian samples obtained in the Juntura Basin provide associations, the significance of which may be pointed out from the study along the Columbia River. Some of the same type of communities which occupied similar habitats may be recognized.

POND-BANK COMMUNITY

The Krebs Ranch local fauna no. 1 has been shown above to be very similar to the sample from UO loc. 2337 of the Black Butte fauna even though they are of different ages. In the Columbia River study it was discovered that some mammals may be considered "index" forms; that is, they are limited to a single community type and thus indicate the presence of this community where they are recovered. The frequent appearance of *Dipoides* (beaver) and *Teleoceras* (amphibious rhinoceras) in collections from the various local faunas in the Drewsey formation, therefore, indicates the presence of the pond-bank community they are peculiar to.

SAVANNA COMMUNITY

The association of *Hipparion* and *Procamelus* in the faunas of the Drewsey formation indicates that the savanna community was present then (Hemphillian). If the frequency of occurrence of the association of these mammals of the savanna community is an indication of the importance of this community in the region at that time, it must be concluded that it was not so common as in the earlier Clarendonian.

GRASSLAND COMMUNITY

In the Drewsey formation faunas (Bartlett Mountain and UO loc. 2360, near Juntura especially) two migrants of particular interest appear. These are *Pliohippus* (single-toed horse) and *Sphenophalos* (antelope). *Pliohippus* is recognized as a migrant since its immediate predecessors are not known from the Northern Great Basin. It has been recognized as an open-grassland form (Shotwell, 1961) and is a prominent member of known open-grassland community samples in other regions (Shotwell, 1955, 1958a). *Sphenophalos* has no predecessors in the Clarendonian Black Butte fauna and is an associate of *Pliohippus* in open-grassland communities. The appearance of *Pliohippus* and *Sphenophalos* in the faunas of the Hemphillian of this region indicates the appearance of this new community. Loc. 2360 of Hemphillian age in the Drewsey formation is found stratigraphically superimposed over the Juntura formation containing the Black Butte fauna of Clarendonian age. In this instance localities producing the Clarendonian fauna are directly below, in the same hillside, those producing Hemphillian fauna (loc. 2360). This locality is of particular interest here although it has produced a small sample. It includes an association of *Pliohippus*, *Sphenophalos*, and *Osteoborus*. It is expected that some forms may have been members of both savanna and open-grassland communities, especially large carnivores such as *Osteoborus*.

The open-grassland community is well represented in the Rome fauna to the southeast of the Juntura Basin. Here *Pliohippus* and *Sphenophalos* are common forms. *Osteoborus* is also present. No quantitative samples are presently available from this fauna. It is interesting to note, however, that the savanna horse *Hipparion* is apparently absent at Rome and other localities to the east in the Northern Great Basin—Juniper Creek Canyon, etc., in the Hemphillian suggesting that the savanna community was not locally represented. Quantitative samples of open-grassland communities of the Hemphillian in the Northern Great Basin are needed before this community can be well understood in this region. We are continually searching Hemphillian beds for sites which may develop into quarries providing this information. Studies of open-grassland communities in other regions where many of the same forms are involved provide a basis for inference as to the make-up of the grassland community of the Northern Great Basin in light of the known fauna of this region. The occurrence of the mammals indicated by such an inference in the Rome fauna and at loc. 2360 in the Juntura Basin as well as numerous other local occurrences strengthens the inference but cannot replace quantitative data.

TABLE 2

QUARRY SAMPLE DATA OTIS BASIN LOCAL FAUNA

	Number of specimens	Number of individuals	Corrected no. sp./ind.
Hypolagus	3	1	7.0
Liodontia	26	2	32.1
Mylagaulus	5	2	6.8
Citellus	8	1	19.5
Pliosaccomys	1	1	2.4
Mammut	4	1	3.9
Hipparion	2	1	1.9
Pliohippus	1	1	1.0
Rhino	1	1	0.8
Prosthennops	6	1	5.5
Procamelus?	9	2	4.5
	66	14	

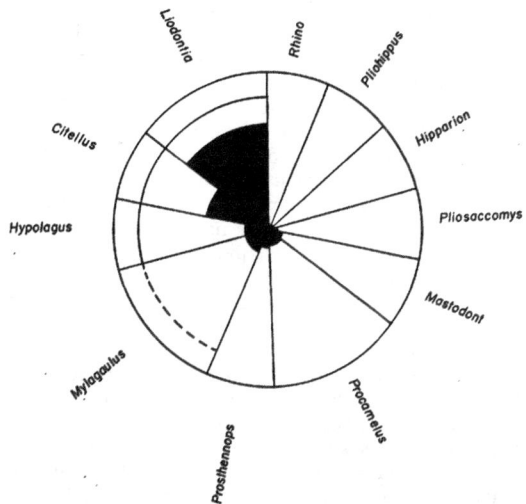

FIG. 4. Faunal analysis diagram from sample at Otis Basin, UO 2347.

We may conclude that the open-grassland community is new to this region in the Hemphillian. It is recognized by the association of *Pliohippus* and *Sphenophalos*, abundant mammals, which are migrants into this region at this time.

BORDER COMMUNITY

Two rodent groups, both containing large forms and usually common, have defied assignment. They often occur in samples but usually as members of distant communities, using the terminology of the approach applied. These two groups are the aplodontid rodents, especially *Liodontia* and the mylagaulids.. Both groups became extinct in the Great Basin after the Hemphillian; however, the aplodontids are still living on the Pacific Coast (*Aplodontia*). A summary of their history may be found in Shotwell (1958b). A sample from the Otis Basin local fauna in the Drewsey formation provides some insight into the community preferences of these rodents. This sample (table 2, fig. 4) is very small but includes both of these rodents as important parts of the fauna. *Liodontia* is a member of the proximal community as are *Citellus* (*Otospermophillus*) and *Hypolagus*. *Mylagaulus* may also be a member of the proximal community or one very close by. Other forms represented from more distant communities are of pond-bank (*Pliosaccomys*), savanna (*Hipparion*), and open-grassland (*Pliohippus*) habitats. It is suggested that this sample represents an environmental situation which is unique but closely adjacent to several others. It is termed here a border community to represent this unique arrangement. Another point to be mentioned is the occurrence of

Mylagaulus in Hemphillian faunas without the presence of *Liodontia*. This is true of many Clarendonian and Hemphillian faunas of the Great Basin. This suggests that these rodents are on occasion members of different types of border communities. Apparently this is not the case in Barstovian faunas of the Northern Great Basin where they usually occur together. The variety of border communities undoubtedly increases as the number of kinds of communities increases. That is to say, with the appearance of new communities (viz. open grassland), more possibilities are present for diversity of border situations; thus we may expect at least grassland, savanna, and woodland border communities. Otis Basin probably represents one of these. These border communities are probably frequently mixed in our samples with those of one of the adjacent habitats and thus may not be readily identified. Such samples as Otis Basin may then be rare and seldom as illuminating as we might like. Their presence must be noted when possible in order to understand the changes of fauna which involve members peculiar to these communities.

COMMUNITY ASSIGNMENTS

DREWSEY

Assignment is based on samples from the Juntura basin except where indicated.

Savanna	Grassland	Pond-bank	Border	Unassigned
*Hipparion	*Pliohippus	*Dipoides	Mylagaulus	Amebelodon
*Procamelus	*Sphenophalos	*Hypolagus	Liodontia	Microtoscoptes
*Osteoborus	*Osteoborus	*Pliosaccomys	Citellus	Hystricops
*Megatylopus	*?Prosthennops	*Teleoceras	Hypolagus	Pliotaxidea
*Mammut		*Prosthennops		

*—Other basins, ?—questionable assignment from available evidence.

The fossil mammals of the Drewsey formation indicate the presence of grassland, savanna, and pond-bank communities. At least one border community is also represented. There is no indication in the unassigned species that a woodland community was present. The open-grassland community makes its first appearance in the Juntura Basin in the Drewsey formation of the Hemphillian. It is unfortunate that we do not have better samples of the open-grassland community. Better samples would provide a more complete list of the members and thus aid in a study of the origin of the various members. Quantitative aspects of the community would also be evident in better samples. The available data only indicate the abundant members of the community. The most important members of the community, *Pliohippus* (single-toed horse) and *Sphenophalos* (antelope), are migrants to the Northern Great Basin in the Hemphillian. *Osteoborus* (hyaenoid dog) is previously a resident of the Northern Great Basin in savanna communities. This is also true for other probable members of the grassland community. Members of the new community then include both migrants and residents. Residents which appear in the new community apparently occupy both savanna and grassland communities.

SUMMARY OF COMMUNITY HISTORY IN THE JUNTURA BASIN AND ADJACENT AREAS

WOODLAND

Barstovian faunas of the Northern Great Basin include significant numbers of browsing horses and artiodactyls, probably woodland forms. Very few of these appear in the subsequent Clarendonian Black Butte fauna. A single species of oreodont and a tapir are the only possible occupants of such a habitat. By the Hemphillian (Drewsey) no woodland elements are present. At present there are no data which might reflect the structure of woodland communities of the Barstovian or Clarendonian.

SAVANNA

The savanna community is well represented in the Clarendonian Black Butte fauna as well as the preceding Barstovian faunas. In the Hemphillian Drewsey faunas the savanna community is also recognized. However, in many Hemphillian sites in the Northern Great Basin the savanna community is apparently not represented as indicated by an absence of the important mammals restricted to this community or its border, *Hipparion* and *Liodontia*. We have made a number of excavations to the east of the Juntura Basin and examined Hemphillian faunas already known from this area. East of the structural trend of the Steens Mountains there are no known occurrences of the important and common savanna elements. Savanna in the Hemphillian is then apparently restricted. Its important members are not known after the Hemphillian.

OPEN GRASSLAND

The open-grassland community is first recognized in the Hemphillian. Its most abundant members are migrants to this region in the Hemphillian.

POND-BANK

The pond-bank community is well represented in the faunas of the Juntura Basin. In the Hemphillian, *Teleoceras* appears as a new member in this community as does *Pliosaccomys*. *Microtoscoptes* may also be a member of this community.

Although some assignments are made to communities on circumstantial evidence, most are based on quantitative studies of samples either from this or earlier work. These results strongly indicate a depletion and eventual loss of woodland communities, a probable loss or at least major reduction of savanna communities at a later date and the appearance of a new community—that of the open grassland. Other communities such as pond-bank appear to be present throughout the sequence studied.

VEGETATIONAL HISTORY

The lower member of the Juntura formation includes massive tuffs. Fossil leaves are relatively common in these beds. All of the mammalian material from the Juntura formation (Black Butte fauna) was collected from the bedded sediments of the upper member of the formation. Chaney (1959) and Gray (in Chaney, 1959) have described the flora of the lower Juntura formation from both leaves and pollen. These floras are found in the valley of Stinkingwater Creek at the western edge of the basin and at Agency Valley Reservoir in the northeastern corner of the basin. Other smaller collections have been made at numerous localities. Chaney considers these floras to be equivalent to the Payette, Succor Creek floras, and refers to them as middle-upper Miocene in age. Mammals from these areas and units are considered to be Barstovian (upper Miocene) by vertebrate paleontologists. These are in turn correlates of the Skull Springs fauna. The Stinkingwater flora represents the vegetation which existed at approximately the same time as the Skull Springs fauna. However, the Skull Springs fauna probably occupied somewhat higher elevations. The nature of the flora at that area can be interpreted from the Stinkingwater flora and Succor Creek flora.

Floral materials are exceedingly rare in the beds producing the mammalian faunas of the Clarendonian. No good flora of this age is known from the Juntura Basin. In fact, paleobotanists do not presently recognize any floras from this entire region which they would assign to the Clarendonian (early Pliocene). However, several floras are known which come from beds which have produced mammals of Clarendonian age. These floras are presently referred to as upper Miocene (Chaney, 1959) or Mio-Pliocene (Axelrod, 1956), although earlier Dorf (1936) considered one of these (lower Idaho) to be of lower Pliocene age. The lower Idaho "formation" has produced a number of Clarendonian mammalian localities since Dorf's work was published. Thus the lower Idaho flora provides a picture of the vegetation of the Clarendonian of this region. The chief argument presented for considering the "upper Miocene" floras (lower Idaho) as Miocene instead of Pliocene, although they occur in units which contain Pliocene mammals, is that they are (the floras) more like Miocene floras. Since no Pliocene floras are now recognized by paleobotanists in this area, it is likely that any early Pliocene flora is going to look Miocene as noted previously by Dorf (1936: 102) even though species occur in it which are restricted to the Pliocene in other regions of North America. The paleobotanist would like the major floral changes to coincide with epoch boundaries. However, since these floral changes are largely ascribed to regional topographic changes (Chaney, 1959: 133) or secular climatic changes, there seems little reason to expect floral changes to coincide with epoch boundaries.

Chaney (1959: 91, 103, 132, 133) has commented critically on our geologic interpretations and age assignments in the Juntura Basin. These remarks were based primarily on our unpublished work in the area as of 1955. Subsequent work in the basin has borne out our earlier impressions as shown in the chapter on the geology of the basin. This work indicates that the basalts of the Stinkingwater mountains are, as we suggested earlier, much younger than the beds containing the Stinkingwater flora. In Pine Creek Valley to the west they overlie the youngest part of the Juntura formation. The flora of the Stinkingwater Creek and Agency Valley Reservoir localities as well as all the less prominent occurrences of plant remains in the basin are found to be in the lower part of the same stratigraphic unit. The vertebrates are higher in the section.

Although the previous remarks may suggest major conflict with paleobotanists, we are in agreement as to the sequence of events and their correlation in vegetational and faunal change. Our differences are in age assignments. These differences are annoying but of little consequence in regional earth-history studies.

Chaney (1959) assigned the plants represented in the Stinkingwater flora (lower Juntura, Barstovian) to three plant associations: lowland, slope, and montane. The lowland association is best represented by leaves. The slope and montane associations are best represented in the pollen and occasionally in the leaves. Chaney notes that the lowland association probably occurred along rivers rather than in swamps in contrast to the Mascall flora of the John Day Basin. This is indicated by the abundance of the water pine, *Glyptostrobus oregonensis* in the Stinkingwater flora, apparently a river-border form, whereas in the Mascall flora of the John Day Basin the swamp cypress *Taxodium* is abundant. The mesic forests of the lowlands also included black oak, alder, sycamore, swamp cottonwood, and elm.

The slopes and benches at middle elevations were occupied by oak woodland and savanna. The occurrence of this association in floras of similar age to the east indicates that this association probably extended at least into western Idaho where it is represented by oaks and other semi-xeric trees and shrubs (Chaney, 1959: 108). This association also includes among others: maple, birch, pine, Oregon grape, spruce, walnut, and a number of herbs and grasses.

The montane association is, as might be expected, poorly represented at Stinkingwater. Chaney indicates fir, maple, alder, birch, beech, spruce, and oak as members of this association.

As noted above, fossil plants are too rare in the upper Juntura formation (Clarendonian) to provide a picture of the vegetation of that time. However, the Lower Idaho flora of Dorf (1936), collected approximately fifty miles northeast of Juntura, can provide an indication of the flora of the Juntura Basin in the early Pliocene. The material studied by Dorf came from several localities which reflect something of the local environmental diversity. Three communities are represented. They are oak-madrone woodland, pine forest, and stream border or riparion communities. The oak-madrone woodland contained oaks, pine, maple, madrone, fir, redwood, and Oregon grape. Plants in the flora associated with pine forest are maple, chinquapin, choke cherry, and hard tack. Those representing stream or riparian environments include willow, horsetail, cattail, ash, alder, and birch. Although grasses are not present in the known fossil flora, they were undoubtedly present. They probably occurred in the form of oak woodland savanna.

This flora as well as others of the same age in the Northern Great Basin may be described as representatives of an impoverished Arcto-Tertiary geoflora (Axelrod, 1950, 1956). They indicate somewhat drier conditions than other floras of the more typical Arcto-Tertiary geoflora, such as the Stinkingwater and Mascall. Rainfall is estimated at from 25 to 30 inches per year, a portion of it in the summer. In contrast, the Stinkingwater flora demanded as much as 40–50 inches of rain.

Hemphillian floras are also rare in the Northern Great Basin. We have not found any fossil leaves in the Drewsey formation in the Juntura Basin. Knowledge of Hemphillian (mid-Pliocene) vegetation is dependent largely on the Alturas (Axelrod, 1944) and Deschutes (Chaney, 1938) floras along the western borders of the Northern Great Basin. Chaney (1944) and Axelrod (1948, 1950) have emphasized the drier character of these floras in contrast to those preceding them. They have also noted that these floras are very much like those of modern stream-bank communities in this region. Forest species of the Clarendonian (early Pliocene) are absent. The vegetation of this period of time provided open environments with widespread grassland in the lowlands. Broad-leaved species were restricted largely to stream-bank communities and higher elevations. Rainfall was about 15–17 inches (Axelrod, 1950).

Although we cannot describe the vegetation of the Pliocene of the Juntura Basin from local floras within the basin, we can be reasonably sure of their character from floras of adjacent areas within the Northern Great Basin. The similarity of the Miocene flora of Stinkingwater which is in the Juntura Basin to the Miocene floras from these same adjacent areas affirms the probability that this is a reasonable inference.

In a recent article Hibbard (1960) criticizes the use of paleobotanical evidence in the interpretation of past climates, especially temperatures. His criticism is in part directed towards the works of Chaney and Axelrod. Since I have relied extensively on their conclusions in this paper it is appropriate that I explain why I still feel the information is reliable.

Hibbard states (1960: 5) "The interpretations of these climates (Late Cenozoic) have been based chiefly upon an extrapolation of the present climate, local environments, habitats, and distributions of the present flora and fauna. From such an approach we have read into the past too much of the present climate. It is more reasonable to begin with the past and work toward the present." On the next page (p. 6) he states how this is to be accomplished, "In the analysis of a flora or fauna it is assumed, unless there is evidence to the contrary, that extinct plants and animals had similar environmental requirements to those of their still-living relatives."

It appears then that Hibbard changed his mind and decided that the present is a key to the past. He assumes in his later statement that there has been no evolution of new morphologic adaptations of the organisms analyzed. Apparently what he means is that in large reptiles, which he is concerned with at this point in his discussion, minimum temperature resistance was probably the same in the early Pliocene as now. In this he is probably correct. The means of analysis he describes, however, is much less critical than that employed by paleobotanists. It should be pointed out that Chaney and his students (Axelrod, Dorf, etc.) use a means of analysis which has three salient features: (1) distribution and habitat requirements of living *morphologically* equivalent species; (2) functional morphology of the preserved plant organs, leaves for the most part; (3) plant associations. These are all more limiting than the statement of analysis presented by Hibbard.

Furthermore, Loverage and Williams (1957: 249) in their discussion of the habitat of the living *Geochelone pardalis babcocki*, the genus referred to by Hibbard, indicate that it occurs in Africa from sea level to 9–10,000 foot elevation in coastal plain, upland savanna, sandy, and thornbrush steppe habitats. No plant association used by paleobotanists has anywhere near this broad a limit to its habitat requirements.

Hibbard's entire discussion is concerned with the Great Plains, yet the works of Chaney and Axelrod are confined almost entirely to Western North America. The climate of the Great Plains in the late Tertiary is quite different from that of the Great Basin of Western North America (see Hibbard's fig. 2) where much of the work he disagrees with was carried out. Hibbard relies heavily on the distribution of large reptiles, particularly crocodilians and turtles to point out his disagreement. It is interesting to note that in the Great Basin, for instance, crocodilians are not known after the Oligocene and large turtles occur only as late as early Pliocene. These large reptiles, used as indications of subtropical temperatures by Hibbard, are lost much earlier in the Great Basin than in the Great Plains. Their distribution changes thus agree well with the interpretations of climate proposed by paleobotanists. It is my conclusion that Hibbard's criticism provides no basis for modification of the conclusions I have used from these works.

CONCLUSIONS

In the Pliocene sequence of faunas of the Juntura Basin, the first appearance, depletion, and loss of mammalian communities coincides with similar changes in the vegetation. Mammals new to the fauna of this area are found for the most part in new communities with a very few found in established communities. To this extent faunal change is a reflection of community change. No instance of replacement of a resident form by a migrant is recognized or suggested by the data. Mammals lost in the sequence studied in the Juntura Basin were apparently members of Clarendonian communities no longer present in the Hemphillian. Changes in the occurrence of phyletic lines of mammals represented in this sequence are reflections of changes in the community types represented and are thus attributed to environmental factors.

Evolution is apparent only in a few instances in the sequence (beavers, mylagaulid rodents, possibly squirrels, etc.) and in these cases the ancestor and descendant forms are assigned to the same community type in successive steps of the sequence. These assignments are based on the nature of the associations in which they are found and are not based on taxonomic affinities. These changes are described in detail in subsequent chapters of this report.

In the Juntura Basin sequence environmental change appears as the major force in the faunal changes noted. These faunal changes reflect the types of mammalian communities present which in turn reflect environmental changes. Evolution occurred only within communities and did not provide members of new communities from former resident phyletic lines by the evolution of new adaptive types. New adaptive types appear as migrants from regions in which their community type, new to the Northern Great Basin, had already existed for some time.

REFERENCES

AXELROD, D. I. 1944. Pliocene floras of California and Oregon. *Carn. Inst. Wash. Cont. Paleo.* **553** (10) : 263–284.
—— 1948. Climate and evolution in western North America during middle Pliocene time. *Evolution* **2** (2) : 127–144.
—— 1950. Evolution of desert vegetation in western North America. *Carn. Inst. Wash. Contr. Paleo.* **590**: 215–306.
—— 1956. Mio-Pliocene floras from west-central Nevada. *Univ. Calif. Pub. Geol. Sci.* **33**: 232–270.
AXELROD, D. I., and W. S. TING. 1960. Late Pliocene floras east of the Sierra Nevada. *Univ. Calif. Pub. Geol. Sci.* **39** (1) : 1–118.
BRODKORB, P. 1961. Birds from the Pliocene of Juntura, Oregon. *Quart. Jour. Fla. Acad. Sci.* **24** (3) : 169–184.
CHANEY, R. W. 1938. The Deschutes flora of eastern Oregon. *Carn. Inst. Wash. Cont. Paleo.* **476** (4) : 185–216.
—— 1959. Miocene floras of the Columbia plateau. *Carn. Inst. Wash. Pub.* **617** (1) : 1–134.
DORF, E. 1936. A late Tertiary flora from southwestern Idaho. *Carn. Inst. Wash. Contr. Paleo.* **476** (2) : 73–124.
GAZIN, C. L. 1932. A Miocene mammalian fauna from Southeastern Oregon. *Carn. Inst. Wash. Contr. Paleo.* **418** (3) : 37–86.
HIBBARD, C. W. 1949. Techniques of collecting micro vertebrate fossils. *Cont. Mus. Paleo. Univ. Mich.* **8** (2) : 7–19.
—— 1960. An interpretation of Pliocene and Pleistocene climates in North America. *Mich. Acad. Rep. 1959–1960*, 5–30.
LOVERIDGE, A., and E. E. WILLIAMS. 1957. Revision of the African tortoises and turtles of the suborder Cryptodira. *Bull. Mus. Comp. Zool.* **115** (6) : 163–557.
MOORE, B. N. 1937. Nonmetallic mineral resources of eastern Oregon. *U. S. Geol. Surv. Bull.* **875**: 1–180.
SHOTWELL, J. A. 1955. An approach to the paleoecology of mammals. *Ecol.* **36** (2) : 327–337.
—— 1958a. Intercommunity relationships in Hemphillian (Mid-Pliocene) mammals. *Ecol.* **39** (2) : 271–282.
—— 1958b. Evolution and biogeography of the aplodontid and mylagaulid rodents. *Evol.* **12** (4) : 451–484.
—— 1961. Late Tertiary biogeography of horses in the Northern Great Basin. *Jour. Paleo.* **35** (1) : 203–217.
WILSON, R. W. 1960. Early Miocene rodents and insectivores from northeastern Colorado. *Univ. Kan. Pal. Cont. Vert.* **7**: 1–92.

2. GENERAL GEOLOGY OF THE NORTHERN JUNTURA BASIN

R. G. BOWEN, W. L. GRAY, AND D. C. GREGORY

University of Oregon

INTRODUCTION

The Juntura Basin in southeast Oregon is the site of deposition of a series of late Tertiary sediments. The sequence exposed consists of Miocene and Pliocene rocks which have produced a number of faunas and floras. Geologic studies in this basin have been aimed primarily at providing stratigraphic controls for the paleontologic work carried out concurrently with the studies described in this chapter. Future work will include detailed studies of the sedimentary units, especially their geometry and environment of deposition. Some aspects of the origin of the igneous rocks are discussed here.

The southern limits of the Juntura Basin are as yet unknown. The northern Juntura Basin is bisected east-west by U. S. Highway 20 and north-south by the Malheur-Harney County line.

The topography has a relief of 2,000 feet with the lowest elevation at 2,700 feet. Junipers grow on the basaltic uplands and hills above 4,000 feet. The climate is semi-arid with much of the precipitation in the form of late summer thundershowers. A weather station, formerly maintained at Beulah near Agency Valley Reservoir, recorded an average rainfall of about ten inches. Most of the basin, however, receives somewhat less precipitation. Temperatures in the late winter are often well below zero and seldom above freezing. Summer temperatures often exceed 100°. Agriculture in the area depends largely on stored water captured from streams draining the mountains to the north. Crops raised are primarily stock feeds, largely grasses for the thousands of head of cattle which are wintered off the range.

Prior to the work reported here, geologic work of only reconnaissance nature was carried out in this area. Russell (1903) investigated the possibility of artesian water in Otis Basin, and Moore (1937) investigated the economic possibilities of the diatomite in Otis Basin. Moore (1937) published a geologic map of Otis Basin. No other maps of this or adjacent areas have been published. The Harney Basin to the west was mapped by Piper *et al.* (1939). The present report is thus the first attempt to delineate the rocks exposed and their relationships.

BASEMENT COMPLEX

The sedimentary rocks of the Juntura Basin rest on a complex of igneous rocks. These extrusive volcanics are exposed in the eastern portion of the basin and extensively in the Malheur gorge east of the basin. At no point in the basin or in the Malheur gorge is the base of this extensive series exposed. In the Juntura Basin this basement complex includes a prominent welded tuff, an olivine basalt and a dark aphanitic basalt with interbedded tuffs and agglomerate beds. The welded tuff is in the lowermost part of the exposed section. The basalts overlie this welded tuff. On the east side of the north fork of the Malheur River a northwesterly trending fault scarp exposes a thickness of approximately 1,700 feet of the basement complex.

WELDED TUFF

The welded tuff is very distinctive and there is no danger of mistaking it for younger welded tuffs locally present. Distinguishing it from the other welded tuffs is its intense silicification by chalcedony. Even miarolitic cavities are partially filled with chalcedony. A weathered brownish-red color is typical although in places it is green. On a fresh surface the color varies from greenish gray to purplish gray. The thickness varies from 100 to 200 feet and averages about 125 feet. The base, where exposed, overlies a light gray tuff which grades up into a very dark-colored baked tuff. The base of the welded tuff resembles a black obsidian and has spherulites 2 cm. to 4 cm. in diameter. A thin section of the rock shows, however, that it is composed of welded glass shards. This obsidianlike bed grades upward into welded tuff which is grayish purple in color and which contains about 30 per cent unfilled vesicles. These vesicles are elongated parallel to the base of the welded tuff and are up to 7 cm. in maximum diameter. They decrease in size and number upward in the section until they are completely or nearly absent.

The massive zone of the welded tuff resembles a pinkish red rhyolite flow with phenocrysts of sanidine sometimes present. This massive zone grades upward into a zone of lithophysae 3 mm. to 2 cm. in diameter. The majority of the vesicles in the upper portion are completely or partially filled with cristobalite, although some are devoid of any filling. The lithophysae zone, in turn, grades upward into an uppermost pink porous tuff which has a platy fracture. In it pieces of pumice up to 2 cm. in diameter can be seen in the hand specimen and vesicles are completely lacking.

The section described above, although typical of the welded tuff, varies at some exposures. The "obsidian" bed is completely lacking in several localities where the base is exposed. The unfilled vesicles and lithophysae are not always present and in several places one occurs without the presence of the other.

Glass shards make up about 94 per cent of the rock. The glass has a reddish color in thin section and is devitrified around the shards. The recrystallization

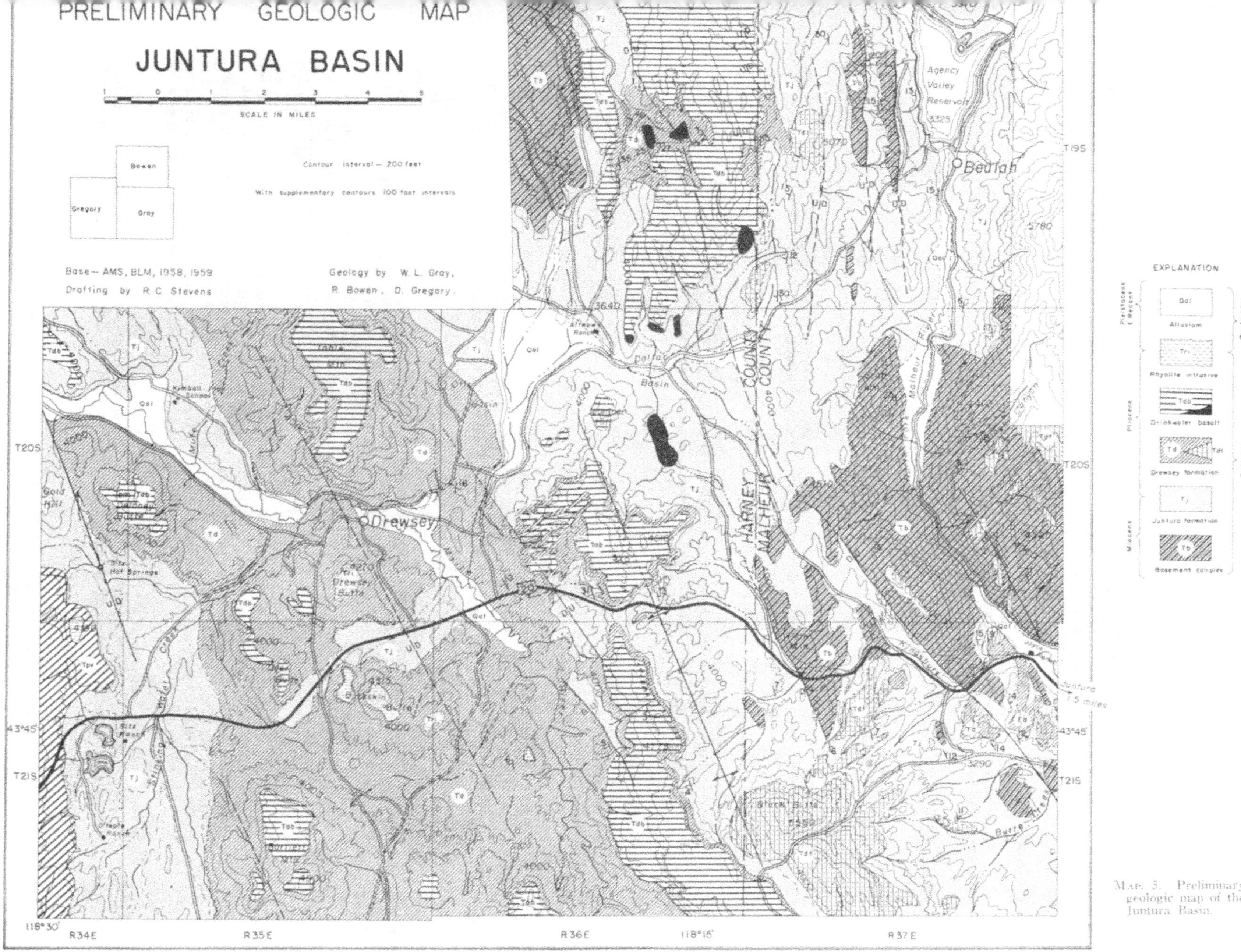
PRELIMINARY GEOLOGIC MAP
JUNTURA BASIN
SCALE IN MILES
Contour interval — 200 feet
With supplementary contours 100 foot intervals
Bowen
Gregory
Gray
Base — AMS, BLM, 1958, 1959
Drafting by R. C. Stevens
Geology by W. L. Gray, R. Bowen, D. Gregory.
Agency Valley Reservoir
3325
Beulah
T19S
T20S
T21S
43°45'
118°30'
118°15'
R34E
R35E
R36E
R37E
Drewsey
HARNEY COUNTY
MALHEUR COUNTY
Basin
Juntura 7.5 miles
Gold Hill
Drewsey Butte
Black Butte
3290
3640
5780
EXPLANATION
Qal
Alluvium
Tri
Rhyolite intrusive
Tdb
Drinkwater basalt
Td
Tdt
Drewsey formation
Tj
Juntura formation
Tb
Basement complex
Pleistocene & Recent
Pliocene
Miocene
Quaternary
Tertiary

Map 5. Preliminary geologic map of the Juntura Basin.

of the glass is in the form of small crystals which grow outward from the surface of the shards. Gilbert (1938) noted similar recrystallization in his study of a welded tuff in the Bishop area of eastern California. He suggests that the product of recrystallization was probably potash feldspar or possibly sanidine and tridymite. A reaction rim of similar material occurs around some of the oligoclase crystals. Devitrification is more pronounced in the upper portion of the welded tuff than in the lower. In the "obsidian" at the base of the member, devitrification was entirely lacking in all except the spherulites.

No consistent increase in flattening of the glass shards toward the base of the section was noted, as reported of welded tuffs from other areas. The porous and less dense tuff at the very top of the section does, however, have unflattened pumice vesicles as well as glass shards. Near the center of the section the shards are somewhat elongated parallel to the base, which along with the elliptical vesicles previously mentioned, suggests that some movement took place after deposition.

The phenocrysts in the welded tuff, which make up about 4 per cent of the rock, are composed of oligoclase, sanidine, and quartz. The sanidine is most abundant, increasing in amount near the top of the section. The phenocrysts average about 1.0 mm. in diameter and both the oligoclase and sanidine show resorption. The resorption is more pronounced in the sanidine which originally had good euhedral outline. Sometime after welding occurred, this rock was intensely fractured and channels were opened for the silica-bearing solutions that permeated it.

Rock fragments usually make up less than 1 per cent of the rock and average about 0.5 mm. in diameter. These fragments are of more basic rocks and are generally composed of dark glass containing microlites of feldspar. Some of the included fragments are of welded tuff which show more complete flattening of the shards than does the surrounding rock. The glass shards of the welded tuff bend around the included rock fragments and phenocrysts. Nowhere in the section were large lapilli or bombs found in the tuff.

Magnetite dust occurs as anhedral grains throughout the rocks and constitutes less than 1 per cent of the rock. Small red specks, which are believed to be hematite, are possibly the source of the red color of the glass.

The lithophysae in the upper portion range from 3 mm. to 2 cm. in diameter and were found to be composed of cristobalite. Eutaxitic texture is faintly visible on the outer rim of the lithophysae but is lacking in the center. The devitrification of the glass shards which is present throughout the section is also present in the outer rim of the lithophysae.

George (1924) found that there is a relationship between silica composition and the index of refraction of a natural glass. The index of refraction of the glass of the obsidianlike portion of the welded tuff was found to be 1.500. This index applied to George's graph shows a silica composition of 72 per cent, which would indicate that the rock is of a rhyolitic composition.

OLIVINE BASALT

The olivine basalt is composed of coarsely holocrystalline olivine basalt flows with feldspar phenocrysts up to 10 mm. in length. The olivine basalt shows a flow on flow structure; in places it has columnar jointing. In the hand specimen it is dark gray in color and weathers to light brown. In some localities the basalt was found to be highly scoriaceous, some specimens with amygdules of secondary calcite. The flows are thin, most having a thickness of about 30 feet. In some of the flows the feldspars are of much smaller size, 2 to 3 mm. in length. In thin section this rock appears as a holocrystalline olivine basalt with an intergranular to subpoikilitic texture. Labradorite An_{60}[2], with laths averaging 3 mm. in length, makes up 70 per cent of the rock. The majority of the feldspar laths are euhedral and have albite twinning, although some show Carlsbad twinning. The feldspars with Carlsbad twinning generally show zoning. Augite constitutes about 15 per cent of the rock, and olivine about 8 per cent in subhedral to euhedral crystals averaging about 0.4 mm. in size. About 5 per cent of the rock is iddingsite which occurs as an alteration product of the olivine crystals; in places the olivine has altered completely to iddingsite. Accessory minerals are anhedral magnetite grains (2 per cent) and apatite (less than 1 per cent).

APHANITIC BASALT

Forming the main mass of Scab Mountain is a black aphanitic basalt and some interbeds of tuff and agglomerate. At the base is the welded-tuff bed. The aphanitic basalt in outcrop can sometimes be distinguished as individual flows, but usually is massive forming rounded knolls. The rock breaks with a conchoidal fracture into small angular blocks and plates, of about two by three by four inches. As these blocks break away from the outcrop they form long sinuous talus streams, five to ten feet wide, which can be traced for twenty-five to fifty yards down the slopes before they are covered by vegetation. Hand specimens are most commonly black and aphanitic with occasional porphyritic varieties. Upon weathering, the rocks turn dark gray to dark red. Most of the hand specimens are so fine grained as to appear glassy, but some are coarse enough to have feldspar laths that can be seen with a hand lens. One specimen contained bytownite phenocrysts up to a centimeter in length.

The interbedded tuffs are generally light to dark gray in color. The tuffs often show good bedding with strata 1 to 3 inches in thickness. Fragments of pumice up to 1 cm. in diameter, and feldspar crystals up to 2 mm. can be seen in the hand specimen of most of the tuffs. Volcanic agglomerate occurs to the north of

Pete Mountain. The agglomerate has a dark gray matrix of tuff with imbedded pieces of pumice up to 5 cm. in diameter. Blocks of dark aphanitic scoriaceous basalt up to 3 feet in diameter are imbedded in this matrix.

A microscopic examination of a typical sample of the aphanitic basalt cropping out on the north end of Scab Mountain shows a hyalo-ophitic texture with about 40 per cent crystalline material set in a matrix of black opaque glass. Labradorite laths ranging in length from about 0.04 mm. to 0.5 mm. with a modal length of about 0.10 mm. constitute 65 per cent of the crystalline material. A pyroxene in equidimensional grains of the same size range accounts for the rest of the crystalline material. The optical constants of the pyroxene are difficult to determine owing to the small size of the grains, but they generally appear to fit the properties of pigeonite.

About 10 per cent of the glass is a green transparent variety with an index of refraction greater than that of Canada balsam. The rest of the glass is black and opaque, probably owing to occult magnetite included within its structure.

AGE AND CORRELATION

The lower member of the overlying Juntura formation is of at least Barstovian age. This minimal date is the only direct evidence of the age of the basement complex. To the east the Owyhee basalts and to the northwest the Strawberry volcanics present possible correlates. They are both complex volcanic sequences, primarily basalts; however, their relationship to the basement in the Juntura Basin is not clear at this time. All three units apparently occupy similar stratigraphic positions.

JUNTURA FORMATION

The Juntura formation takes its name from the town of that name in Malheur County, Oregon. No single section reveals all three members of this diverse unit. A typical section of the lower member is that exposed north of Scab Mountain, of the middle member, Juniper Hill and of the upper member, the vicinity of quarry 3 south of Pete Mountain. A composite section of these appears in figure 5.

The Juntura formation covers approximately one-half of the area mapped, diatomaceous beds making up the greater part. The most extensive occurrence is in the eastern portion of the basin with lesser exposures along the valley of Stinkingwater Creek.

The Juntura formation is approximately 1,250 feet thick. North of Scab Mountain about 400 feet of tuffaceous and agglomeritic beds capped by a flow of palagonite basalt are exposed. At Juniper Hill 450 feet of diatomite were measured; however, the base of the diatomite was not exposed. To the south of Pete Mountain the upper tuffaceous beds are 400 feet thick.

The Juntura formation disconformably overlies the basement complex although there may also be a slight angular unconformity. No consistent difference of dip was obtained between the sediments and the underlying basaltic flows. There was, however, a period of deep erosion of the older basalts before deposition of the Juntura formation. Deeply weathered red basalt flows underlie the sediments southeast of Pete Mountain. Pete Mountain, which is surrounded at its southern end by the Juntura formation, was a highland when the lacustrine sediments were deposited. The eleva-

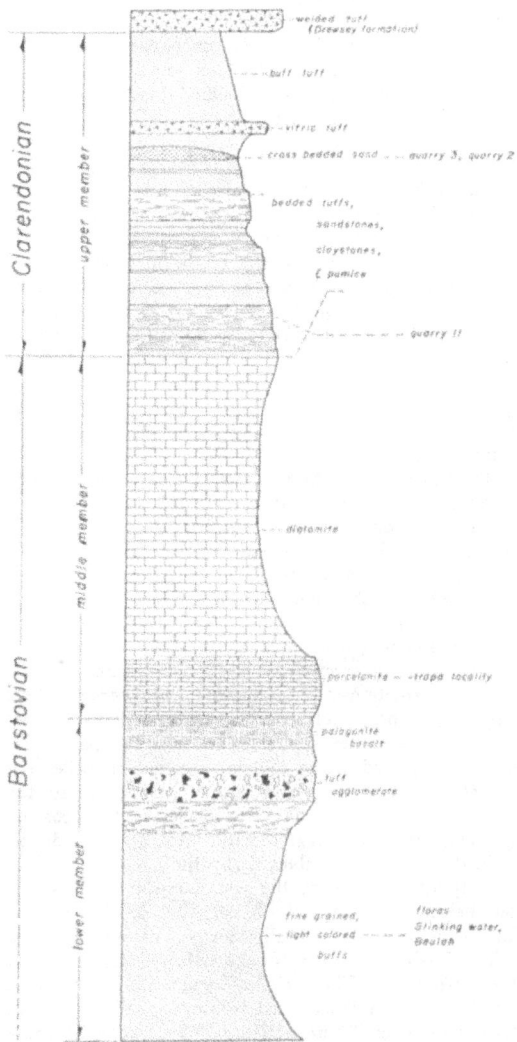

Fig. 5. Composite measured section of the Juntura formation. Scale: 1″ = 200′.

tion of Pete Mountain is 300 or 400 feet above the top of the Juntura formation.

The basement basalt crops out through the Juntura formation east of Black Butte where the lake beds have been nearly eroded away. Exposures of the underlying basalt occur in a series of low northwest trending ridges with the lacustrine sediments filling in the valleys between the ridges.

The Juntura formation is predominantly light colored; diatomaceous beds are white to cream in color and the upper beds vary in color from cream through shades of tan to brown. The beds are well stratified, except in areas where the purer diatomite occurs. The sediments are generally poorly consolidated except for occasional thin beds of tuffaceous sandstone near the top of the series and beds of silicified diatomite near the base of the middle member.

LOWER MEMBER

The unit as exposed in the west side of the valley of the North Fork Malheur River north of Scab Mountain consists predominantly of fine-grained unconsolidated ash. In the upper part of the member, beds of coarse-grained tuffaceous agglomerate and flows of palagonite basalt form distinctive north striking ridges.

The ash beds occur as rounded hillsides that show little dissection because of their very loose, permeable nature. Owing to this feature, they can be distinguished readily in the field from the more consolidated tuffs and tuffaceous agglomerates which erode into a rough irregular surface.

The ash is usually fine grained, but at some horizons it attains the size of lapilli. The color is variable depending upon the state of weathering and size of the individual grains. In general the finer-grained ash is light gray to buff yellow when weathered and dark brown when fresh. The coarser-grained ash is usually a darker shade of the same colors. No microscopic examination was made of these ashes.

Higher in the section of the lower member of the Juntura formation is the tuffaceous agglomerate. This agglomerate occurs in beds ranging from 10 to 25 feet thick and generally seems to underlie the palagonite basalt. One bed of the agglomerate that underlies the highest flow of palagonite basalt can be traced for several miles along strike. It grades from a coarse-grained tuffaceous agglomerate into a hard, dense, welded tuff that resembles a rhyolite.

A typical sample of the agglomeratic phase crops out below the palagonite basalt just to the east of Dollar Basin. The overall color is light gray with a faint bluish cast when fresh, which weathers to a brownish gray. The rock is characterized by a large number of equidimensional white to yellow pumice fragments of lapilli size. The pumice fragments show no flattening. The groundmass consists of fine-grained glass shards with some intermixed feldspar fragments. In general the rock is very porous. Under the micro-

scope the pumice fragments show a sublinear arrangement with their long axis in approximately the same direction. The long dimensions of the pumice grains range between 0.05 cm. and 1 cm., with an average size of about 0.5 cm. The matrix of the rock is made up of both brown and clear glass; the brown glass has an index of refraction greater than that of Canada balsam. The brown glass is probably tachylyte, whereas the clear glass is the more normal acidic variety. Plagioclase feldspar fragments ranging from 0.2 to 1 mm. in length with a mode of about 0.5 mm. make up about 5 per cent of the matrix. On the basis of its nearly parallel extinction and the fineness of the twinning laminae this feldspar is believed to be oligoclase. In one slide there is a pyroxene crystal that has been embayed; a rim of plagioclase crystals penetrates nearly the entire thickness of the pyroxene crystal. This rock is classified as a pumiceous lapilli tuff.

Along the strike of this bed the tuff grades into a rock that is compositionally very similiar, the primary difference is in the extreme flattening and elongation of the pumice fragments. This sample is harder and denser and much of the pore space has been eliminated by the compaction of the rock. Under the microscope the effect of this compaction shows in the state of the glass shards which have been flattened and tend to align themselves with the pumice fragments. The flattened pumice fragments tend to be darker than the matrix and show signs of alteration by their slight devitrification. This rock is classified as a vitric tuff that has been partially welded.

The end product of the welding of the glass shards is the true ignimbrite or welded tuff. In outcrop these rocks resemble a rhyolite more than a tuff. In this area the welded tuffs sometime occur along strike with a normal tuff. At a distance the welded tuff can be distinguished by its tendency to form more prominent outcrops than the unwelded facies. As the tuffaceous agglomerate is traced along strike it changes from a subdued ridge, barely differentiated from the surrounding soil, to a ledge that frequently projects 20 to 50 feet up dip from the beds enclosing it. The outcrops of the welded tuff are usually distinguished by a columnar jointing pattern, a vitreous luster on many of the weathered surfaces and a ringing sound when struck with a hammer.

A sample found along strike from the blue agglomerate is typical of the welded tuff found in this area. This rock is about the same color as the two examples previously described. It differs in its other characteristics. The intensity of the welding is illustrated by the extent of the flattening and coalescing of the pumice fragments, which have eliminated all the pore space and produced a flowlike structure in the rock. When fresh, the rock has a dull earthly luster but on exposure takes on a vitreous luster which from a distance gives it the appearance of obsidian. Under the microscope the rock is seen to be composed of about 20 per cent grains set in a matrix of glass shards. Half of the grains are crystals and angular fragments of oligoclase ranging in size from 0.05 mm. to 1 mm.; the mode is about 0.3 mm. With the exception of a few small subhedral crystals of a green pyroxene, the rest of the grains are angular glass fragments, many of which have become devitrified. The matrix is made up of glass shards and pumice fragments. The pumice fragments are distinguishable from the glass shards only by their lighter color and larger size, for they have been so flattened that they wrap around more solid particles as do the glass shards. In this rock the welding has been so complete that the individual shards have coalesced; their former edges are represented only by color variations. The index of refraction of the glass is generally less than that of Canada balsam. The long axis of the oligoclase crystals and the general trend of the shards and pumice fragments are parallel to the margins of the flow. This rock is classified as a welded crystal-vitric tuff.

Separating the lower and middle members of the Juntura formation is a flow of palagonite basalt. This basalt occurs in thin flows east of Dollar Basin where it caps most of the north-south trending ridges. These flows range from 20 to 30 feet thick, striking north-south and dipping 10° to 25° west. At the outcrop they are readily differentiated from the basalts of the basement complex by their coarser grain, a tendency to weather into rounded blocks, and by the fact that they do not form the long talus slopes so common to the older basalt. The bottoms of these flows are not apparent but their red scoriaceous tops can frequently be seen where the cover has been stripped off.

A typical sample of this rock from the ridge east of Dollar Basin has an over-all dark gray-brown cast when fresh and weathers to a light brown. Megascopically the rock might be classified as a diabase because of its ophitic texture. The rock is characterized by a high percentage of pale yellow to brown glass. Aside from the glass the only mineral that can be seen with a hand lens is feldspar in laths up to a millimeter in length. Under the microscope this rock is hypocrystalline. It is made up of about 25 per cent glass, most of which has been altered to palagonite. It has an ophitic texture with labradorite laths ranging from 0.01 mm. to 1 mm. in length enclosed in pigeonite. The only other important constituent is ilmenite which makes up about 7 per cent of the rock. There is a trace of olivine and iddingsite.

The palagonite in the basalt was possibly the result of extrusion of the basalt into the lake in which the diatomite was deposited. No pillow structure was found in these basalt flows.

MIDDLE MEMBER

The middle member of the Juntura formation is characterized by extensive deposits of diatomite and associated porcelaneous and opaline rocks. The thickest deposits of diatomite are found in Drinkwater Mountain, and to the north in Otis Basin.

To the east of Drinkwater Pass the lower portion of the diatomite series is exposed. Here the sediments are highly silicified and the better grade of diatomite is lacking. Bramlette (1946) refers to similar beds of silicified diatomite in the Monterey formation as porcelanite. Porcelanite resembles chert, but is not as hard and dense. These silicified rocks are generally yellow or light brown in color, have a splintery fracture, a dull luster, and in some instances resemble unglazed porcelain. The porcelanite tends to break into small fragments on exposure to air. Along U. S. Highway 20, just east of Drinkwater Pass, are a number of small rounded hills 20 to 25 feet high which are composed predominantly of small broken fragments of porcelanite. The porcelanite often shows good bedding 1 to 3 inches in thickness. It is more pronounced where the diatomite has not been so completely silicified.

The diatomite has been intensely silicified locally. These zones are represented in the southern part of local basins and in the area to the south by small hills of porcelaneous rock having cores of opaline chert. These hills range in size from small knobs no higher than a man to some a hundred feet high. Occasionally they are associated with igneous rocks but more commonly there are no visible adjacent igneous rocks.

Porcelanite occurs near Juniper Hill where the diatomite is in contact with igneous rocks which have intruded it. At the contact with the intrusive rock, the diatomite is altered to a brown hard silica rock which grades outward to pure diatomite. These zones of siliceous diatomite are generally 10 to 20 feet thick and roughly parallel the contact of the intrusive rock. The heat and gases from the intrusive rock probably altered the diatomite to porcelanite with the alteration most pronounced at the contact. Porcelanite also occurs north of Buckskin Butte in the western portion of the basin.

The diatomite is white, massive, and relatively pure from Drinkwater Pass northward into the Otis Basin. The absence of volcanic ash in much of the diatomite indicates that the deposit was formed under rather calm conditions. In the upper beds the color of the diatomite changes from white to shades of yellow. Here it is mixed with ash and tuffaceous material. The upper beds of diatomite often show good bedding which is apparently a result of the mixture of ash with the diatomite. Black, well-bedded shale crops out within the pure diatomite in a number of places along the Drewsey cut-off road. A thin section shows this shale to be composed predominantly of diatoms mixed with carbonaceous material.

UPPER MEMBER

The upper tuffaceous beds of the Juntura formation are best exposed to the south and west of Pete Mountain. Since the sediments are poorly consolidated and easily eroded, they crop out only in areas where the welded tuff of the Drewsey formation has protected them from erosion. Where the welded tuff is absent only the diatomaceous beds remain or the sediments have been completely eroded away. This tuffaceous material erodes into slopes with an angle of 15 to 30 degrees. Although the sediments are easily weathered, the covering of alluvium is generally only 3 to 6 inches deep. This lack of alluvium is largely due to deflation by the wind, which carries the fine-grained weathered material from the face of the slopes and in some places has deposited it in sand dunes upon the overlying welded tuff or on the lee side of exposures.

When the thin covering of alluvium is scraped away, the underlying sediments are seen to be thinly and regularly bedded, with laminae ranging from 0.2 to 2 inches in thickness.

Stream-channel deposits cut these tuffaceous sediments in several localities. These fluvial deposits are partially consolidated and show graded bedding from fine-grained conglomerate at the base to sandstone with an average grain size of about 0.3 mm. at the top of the beds. Occasionally these deposits are cross-bedded. The beds include grains of metamorphic, granitic, and hornblende basaltic rocks, none of which crop out within the basin. These sediments probably come from the north where metamorphic and granitic rocks are exposed.

Glass sand beds a few inches to several feet thick occur in both the diatomaceous and the upper tuffaceous beds. A bed of bluish-gray pumice sand forms a ledge 14 feet thick near the top of the series of tuffaceous sediments south of Pete Mountain. This ledge, which maintains a uniform thickness, can be traced for more than a mile. The sand, which shows cross-bedding in some places, is composed almost entirely of platy and angular grains of pumice glass. These tuffs represent an extensive ash fall in the region. A possible explanation for the cross-bedding in this bed is that the sand was concentrated in dunes by the wind.

AGE AND CORRELATION

Floras from the lower member of the Juntura formation (Chaney, 1959) indicate a Barstovian age in the continental Tertiary chronology. A diatom flora is described from the middle member of the Juntura formation by Lohman (in Moore, 1937). Fossil mammals from the upper member of the Juntura formation indicate a Clarendonian age (Shotwell and Russell, chapter 4). This extensive series of sediments then includes beds of late Miocene through Pliocene in age. Fossil mammals from the Poison Creek formation of Idaho, The Dalles formation in north central Oregon, and

the Truckee formation of Nevada indicate that these are all correlates of the upper member of the Juntura formation. Similarity of the lithologic sequence in the Danforth formation to the west of the Juntura Basin suggests that at least part of the Danforth is probably equivalent to the upper Juntura. Floras and faunas from the Mascall formation, Beatty Butte, Skull Springs, and Sucker Creek indicate that these beds are roughly correlates of the lower member of the Juntura formation and possibly the middle member (Chaney, 1959; Downs, 1956).

DREWSEY FORMATION

The Drewsey formation unconformably overlies the Juntura formation and is overlain in some locations by younger basalt flows. The beds are composed predominantly of rhyolitic pumice tuff, agglomeritic mud flows, agglomerates, gravels, grits, sands, silts, tuffs, cinders, volcanic ashes, and basalt flows. Some of this material is reworked from the Juntura formation. The sandstones, mudflows, agglomerates, and tuffs are usually well consolidated and, therefore, resistant to erosion. Other materials are generally of a loose and porous nature. All the beds are highly lenticular which makes it difficult to correlate adjacent sections. In contrast to the Juntura formation which is predominantly fine-grained and light-colored, the Drewsey formation is generally coarser, brown, gray and "dirty-looking." The Drewsey formation lies mainly within a northwest-southeast trending syncline that passes east of the town of Drewsey and along the west flank of Drinkwater Mountain. When sediments were deposited on the flanks of the syncline, a small angular unconformity exists between the Drewsey formation and the underlying Juntura formation. The synclinal axis may have received sediments without much time lapse. The greatest thickness of the Drewsey formation, of over 1,000 feet, occurs in this depression.

A section of 1,050 feet of Drewsey formation was estimated along Mule Creek west of Table Mountain. The base of the Drewsey formation is not exposed here. Remnants of the Drewsey formation overlie a welded tuff on the eastern side of the basin where younger terrace deposits have protected the beds from erosion. On the eastern limb of the anticline, which has its crest at Drinkwater Mountain, the basement complex and the Juntura formation both dip 10 to 15 degrees. The overlying Drewsey formation has a dip of about 8 degrees. The anticlinal folding is also reflected in the later Pliocene basalt flows, which have a dip of 2 to 5 degrees away from the crest of the fold. This may represent an initial dip.

No single exposed section includes all the lithologies of the Drewsey formation. Any type section that we might specify is thus an incomplete one. The section exposed in Table Mountain just north of the town of Drewsey and directly across the Malheur River is designated the type section. At this point there is an

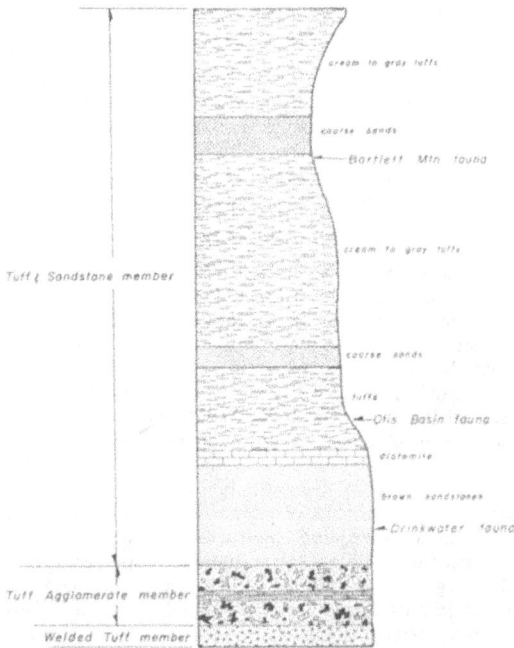

FIG. 6. Composite measured section of the Drewsey formation.
Scale: 1″ = 200′.

exposure of a little over 1,000 feet of section. Some lithologies characteristic of the Drewsey formation in other portions of the basin (viz. welded tuff) do not occur here. We shall refer to additional sections where particular lithologies are best seen.

Since the beds of the Drewsey formation are often lenticular, it is difficult to correlate sections using individual beds. However, several general divisions of the beds seem to be most effective in relating sections. These divisions are described as members in the following discussion.

THE LAPILLI-TUFF, WELDED-TUFF MEMBER

The lowest member of the Drewsey formation consists of a welded tuff in the eastern portion of the basin and a lapilli tuff in the western portion. They occupy similar stratigraphic positions and may, in fact, represent different aspects of the same rock. The best exposures of the welded-tuff bed are north of Butte Creek where erosion has removed the overlying sediments. The color of the bed is generally light gray where it is unwelded and dark gray in the welded portions; however, yellow and red varieties are found at some localities. The bed shows a fairly uniform thickness, averaging 15 feet, over a wide area. The welded portion of the bed usually breaks in a subconchoidal fracture, and on a fresh surface the luster is resinous.

In places the welded tuff exhibits spheroidal weathering. The welded tuff is generally free of lapilli, although, in a few outcrops, fragments up to 2 cm. in diameter were found imbedded in the rock. In all samples of the tuff, whether welded or unwelded, euhedral crystals of sanidine and quartz 1 to 4 mm. in length are to be seen in the rock.

The bed shows gradation both laterally and vertically from unwelded to welded tuff. A good example of vertical gradation in the welding can be seen in the bed above fossil locality UOMNH 2332 (see map 2). Here the base of the tuff is a light gray color and is composed of porous unwelded tuff. It becomes more compact upward, and about 5 feet from the base is dark gray in color and completely welded. The welded portion of the tuff grades into light gray unwelded tuff at about 4 feet from the top of the bed. Roberts (1956) reports similar welding in an ignimbrite in Central Nevada.

Lateral variation in the welding of the tuff is best shown near Drinkwater Pass. At the pass the tuff is completely welded, but, as the bed is traced westward, the welding becomes less pronounced. At the west base of Drinkwater Mountain near Highway 20 it shows no welding at all and is much thicker, about 35 feet thick. As a general rule, where the tuff lacks welding it is much thicker than where it is welded. This would suggest that settling of the bed accompanied the welding. Fenner (1923) described a sand flow in the Valley of Ten Thousand Smokes, which shows evidence of settling as much as 100 feet on cooling and welding.

Three samples of the welded tuff were analyzed microscopically from a locality where the bed shows vertical variation in welding. A sample from the base of the bed, is a light gray vitric tuff. Phenocrysts, which make up 5 per cent of the rock, are composed of quartz (1 per cent) and sanidine (4 per cent). These phenocrysts have an average length of about 2 mm. A few devitrified glass fragments and anhedral magnetite grains are scattered throughout the rock. The groundmass is composed of glass dust which lacks devitrification. No glass shards are present in the rock. A sample from the center of the bed is a dark gray welded tuff with eutaxitic texture. The glass shards are flattened, aligned, and bend around included crystals and fragments. The glass shards show a slight devitrification. Euhedral crystals of sanidine and quartz, in about equal amounts and averaging 1.5 mm. in length, make up about 7 per cent of the rock. Also occurring as grains in the rock are euhedral aegirine-augite crystals (2 per cent), devitrified glass fragments (1 per cent) and included fragments of welded tuff (1 per cent). The only plagioclase feldspar in the rock occurs in an included rock fragment. Iron stains surround some of the grains in the rock, and anhedral magnetite grains are scattered throughout the rock. A sample from the top of the bed in the same locality is a light gray tuff which resembles, in the hand specimen, the tuff at the base of the bed. Its composition is nearly the same as that from the central welded portion of the bed. The chief differences are that in this sample aegirine-augite is not present, the glass is not devitrified, and the shards, though aligned, are only slightly flattened.

The pumice lapilli tuff found in the western Juntura Basin may be a continuation of the welded tuff. North of Bartlett Mountain a thin, reverse-bedded pumice layer underlies the pumice lapilli tuff. It averages one foot in thickness; this same pumice bed, or a very similar one, underlies the welded tuff at Upton Mountain south of the mapped area and immediately north of U. S. Highway 20, four miles west of Juntura. The pumice lapilli tuff occurs at approximately the same stratigraphic position within the Drewsey formation as the welded tuff.

The mapping of definite contacts between the Juntura and Drewsey formations is difficult in the western Juntura Basin. This difficulty is further increased by the great number of small faults in the area. Some sediments of the Drewsey may underlie the pumice lapilli tuff member, for no definite erosional break between the Juntura and Drewsey formations is apparent in the axial, synclinal deposits. The pumice lapilli tuff is buff to tan in color. Lapilli up to three inches in thickness occur throughout. The average thickness of the entire unit is approximately 20 to 30 feet. However, in only one locality is the entire thickness exposed. In most places it has been so eroded and weathered that it can be seen only by digging into the hillside. Microscopically, the pumice layer can be classified as a vitric lapilli tuff. It is composed entirely of glass shards and pumice set in a slightly devitrified groundmass.

TUFF AGGLOMERATE MEMBER

A dissected tuff cone is a prominent feature located three and one-half miles north of the Altnow Ranch. The tuffaceous agglomerates of which it is composed also occur along Mule Creek, on the west flank of Drinkwater Mountain, and to the east of the tuff cone. Near the junction of Cottonwood Creek and the Malheur River the agglomerate has basaltic blocks up to six inches in diameter in a fine-grained matrix of yellow tuff. At other localities the basaltic blocks reach as much as two feet in diameter. Basalt flows on the west flank of Drinkwater Mountain near U. S. Highway 20 are apparently interbedded with the sediments of the tuff-agglomerate member. Other basalt flows at the same stratigraphic horizon occur at Drewsey Butte and Buckskin Butte.

A bed of diatomite about 25 feet thick crops out on the hillside just east of Drinkwater Ranch. The diatomite is poorly bedded and has a high percentage of ash mixed with it. Vertebrate fossils are found weathered out on the hillside at this locality and appear to have come from the diatomite bed. The diatomite is apparently at the same stratigraphic level as the upper agglomerate beds locally present.

The tuff agglomerate member thus includes bedded volcanic mudstone, shales, siltstones, sandstones, tuffs, agglomerates, breccias, and possibly diatomites. It is often 75 to 100 feet thick in the synclinal area and much thicker near apparent tuff cones. The section exposed just east of the Malheur River on the west side of Drinkwater Mountain is typical.

TUFF AND SANDSTONE MEMBER

In the synclinal area, Table Mountain in particular, and north of Bartlett Mountain a thick series of volcanic ash and tuff is a prominent feature of the Drewsey formation. Much of the tuff appears to be redeposited from the Juntura formation. A well-cemented tuffaceous sandstone is a dominant feature and may best be seen on the south side of Table Mountain. The sedimentary member has the widest distribution of the various lithologies of the Drewsey formation. It accounts for the majority of the formation exposed at the type section. Most of the vertebrate fossils collected from the Drewsey formation came from this unit.

Black Butte in the eastern Juntura Basin is composed of a large rhyolite mass which intruded the diatomite of the Juntura formation in the middle Pliocene. This intrusive extends south of the area mapped where it is eventually concealed by the Drinkwater Basalt. The intrusive has the appearance of an old volcano, but it may represent a fissure type of eruption since it is elongated in a north-south direction. Vertical columnar jointing with pillars about 75 feet high can be seen in a canyon on the east side of Black Butte. Erosion of the diatomite on the north end of Black Butte has exposed a thick section of the intrusive. Here the rhyolite can be seen to bend from a vertical dikelike mass into horizontal flows which are interbedded with the Drewsey formation.

Samples of the Black Butte rhyolite intrusive were analyzed in thin section and found to be rhyolite porphyry. It is a bluish-gray to light-gray rock with flow structure. Clear euhedral phenocrysts, which make up about 10 per cent of the rock, are composed of about 7 per cent euhedral sanidine and 3 per cent subhedral to anhedral quartz. A micrographic intergrowth of sanidine and quartz is present in many of the crystals of the rock. Sanidine and quartz both occur as small grains as well as phenocrysts in the groundmass. Magnetite and apatite occur as euhedral accessory minerals.

AGE AND CORRELATION

The Drewsey formation has produced a Hemphillian mammalian fauna. The fauna indicates that this unit may be correlated with the Rattlesnake formation of the John Day Basin to the north. The similarity of the sequence of lithologies in the Drewsey formation suggests that it may also be correlated with the upper part of the Danforth formation in the Harney Basin to the west.

The fauna of the Drewsey formation is discussed in chapter 5 of this report.

DRINKWATER BASALT

A nearly horizontal flow of olivine basalt unconformably caps the sediments in the basin. Table Mountain is capped with a flow of this basalt and, in the south, Drinkwater Ridge is also capped with it. This basalt may have been continuous throughout the area before erosion removed all but the remnants now present. The basalt capping has been broken and displaced by several northwesterly trending faults. These faults show little displacement, the movement in most cases is less than a hundred feet. The relief in this area has been accentuated by stream erosion into the diatomite following fracturing of the basalt cap. The basalts are of greatest thickness in the southern part of the basin and thin to the north. Just north of Drinkwater Pass a thickness of 60 feet of basalt was measured. Farther north at Juniper Hill a small outcrop of basalt about 15 feet thick caps the diatomite. Intrusive basalt crops out in the diatomite of the Juntura formation to the north and east of Juniper Hill in the form of small hills 10 to 25 feet high. The surrounding diatomite has been altered to porcelanite as mentioned previously. Moore (1939) mentioned dikes near this locality which he believed to be feeder dikes for the capping basalt flows. The composition of this intrusive resembles that of the capping basalt flows.

The Drinkwater basalt is found as an individual flow, with the exception of the dikes that probably were the feeders. The thickening and splitting into a number of individual flows near Drewsey and south of Drinkwater Mountain probably indicate some of the sources of this basalt. These flows are usually scoriaceous and blocky on their upper surfaces but sometimes are platy. Most commonly the basalt has a dense aphanitic texture with an occasional elongate feldspar lath set in a black groundmass. The usual color is black when fresh, weathering to a dark gray-brown. When there are numerous large feldspar laths visible, the rock takes on a lighter value. A typical sample of this basalt is dark gray in color. Numerous thin laths of feldspar up to 2 mm. long are visible. With a hand lens many grains of olivine, rimmed with iddingsite, can be seen. A microscopic examination of the rock shows an intergranular texture with pyroxene, olivine, and glass filling the spaces between the aligned crystals of labradorite. Labradorite ($Ab_{40}An_{60}$) crystals ranging in size from 0.3 mm. to 4 mm. with a modal length of about 1 mm. make up nearly 50 per cent of the rock. Pigeonite, the next most abundant mineral, makes up about 22 per cent, occurring in small tabular crystals filling the interstices between the labradorite laths. Ilmenite is unusually abundant in that it forms about 10 per cent of the rock. Olivine in equidimen-

sional grains up to 0.6 mm. in diameter makes up about 5 per cent. Iddingsite is commonly associated with the olivine. The remainder of the rock is glass, most of it black and opaque, but some of it is the yellow tachylitic or palagonitic variety.

The type section is at Drinkwater Pass.

The Drinkwater basalt was deposited on an erosional surface formed by the base leveling of both the Juntura and Drewsey formations. It is thus younger than Hemphillian in age. It has, however, been in position long enough to reflect some faults and to have been almost completely removed in some areas in which it probably originally occurred. It must then occupy, in time, a position somewhere between the later Pliocene and early Pleistocene.

A series of basalt flows in the east side of Stinkingwater Mountain may represent the Drinkwater basalt. These basalts are best exposed on the Pine Creek road, ten miles north of U. S. Highway 20, where they overlie the upper member of the Juntura formation. The overlying basalt series is petrographically similar to the Drinkwater basalt. They are dark gray, often porphyritic, vesicular basalts which weather to shades of brown. They generally contain a minor amount of olivine which is rimmed with iddingsite. The flows individually are thin, averaging fifteen feet in thickness. They are not columnar. As many as six flows were observed along Pine Creek, outside the basin. The surface of some of the flows is oxidized and reddened. On extrusion the flows had an initial dip because bubble-train attitudes are still as nearly vertical as they were at the time of formation. To some extent the flows appear to feather out at the base of Stinkingwater Mountain where they come in contact, with the lower member of the Juntura formation. The basalt is observed in contact, other than fault contact, with an ash bed at only one place on the east side of the mountain. Here, a spheroidal basalt immediately overlies a sedimentary ash bed. At this point the basalt resembles a pillow lava. The few "pillows" present are glassy on the outer rim and appear to be layered.

Microscopic analysis of a sample from the basalts on Stinkingwater Mountain along U. S. Highway 20 reveals a merocrystalline, fine-grained basalt. Subparallel laths of labradorite, An_{58}, are set in a matrix of augite, some of which has been altered to serpentine and glass, which is partially devitrified. Magnetite occurs as an accessory mineral. Another sample of basalt, from Stinkingwater Mountain, which is located near a younger rhyolite in McMullen Pasture, shows a somewhat vesicular, fine-grained basalt with trachytic feldspar laths set in a groundmass of black, magnetite-rich glass. A vein running through the thin section shows alteration. Cinnabar is present in this altered zone.

Chaney (1959) has stated that the Stinkingwater flora was collected from beds which overlie the basalt discussed above. The Stinkingwater flora is found in the lower member of the Juntura formation. Along

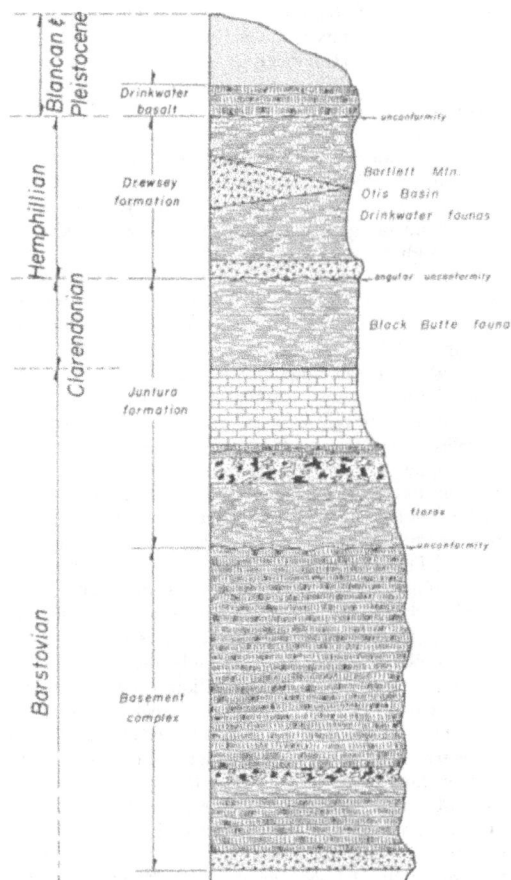

FIG. 7. Stratigraphic position of fossil localities in the Juntura Basin. Scale: 1″ = 400′.

Pine Creek, to the west, these same basalts overlie the upper member of the Juntura formation. These basalts are therefore much younger than he suggests. This does not, however, effect the relative ages of the fossil plants and animals (see fig. 7).

The evidence indicates that these basalts are of Pliocene age, are petrologically similar to the Drinkwater basalt, and for the most part occur at about the same elevations. On the other hand they exhibit multiple flows, which the Drinkwater does not, and they occur at lower elevations on the east flank of Stinkingwater Mountain. These low elevation flows appear to feather out which may indicate, locally, the eastern limits of this unit.

RHYOLITE INTRUSION

A large domelike, steep-sided mass of rhyolitic lava, which occurs at the north end of Stinkingwater Mountain on the west edge of the basin appears to have been squeezed up over the surface of the basalts on Stinkingwater Mountain. This very viscous siliceous lava flowed a short distance into the basin congealed on a steep slope. It is varied in appearance; the most common aspect is its pink color and fine-grained texture. This rhyolitic mass, the one at Drewsey Butte, and one at Buckskin Butte are all in a similar stratigraphic and topographic position.

SUPERFICIAL DEPOSITS

TERRACES

Terrace deposits of sand and gravel occur over the middle Pliocene sediments on the east side of the area to the south and west of Pete Mountain. The elevation of these deposits is about 600 feet above the present elevation of the North Fork Malheur River.

A field examination of these terrace deposits showed the gravels to be well-rounded pebbles and boulder ¼ to 6 inches in diameter. The gravels are composed mainly of massive and scoriaceous basalt. About 5 per cent of the gravels are made up of metamorphic, granitic, and rhyolitic pebbles. These gravels undoubtedly originated to the north where granitic, metamorphic, and rhyolitic rocks are exposed.

Terrace deposits along the west side of the Malheur River west of Drinkwater Mountain are about 400 feet above the present elevation of the river. They are composed of well-rounded cobbles of a more uniform size. These cobbles, which average about one inch in diameter, have no large boulders mixed with them as do the terrace deposits along the North Fork Malheur River. The gravels are composed mainly of basalt with a few metamorphic cobbles mixed with them.

Terraces, representing a period of base leveling by streams entering Otis Basin, are cut into the diatomite in the center and along the margins of the basin. The flat-lying center terrace, located between Otis and Cottonwood Creeks, rises about a hundred feet above the floor of the basin. On the east and west sides of the basin the terraces slope into the center of the basin. A basaltic lag gravel with intermixed fragments of welded tuff covers their surface.

ALLUVIAL FANS

Along the eastern edge of Otis Basin three large alluvial fans have formed where streams draining the ridges to the east have emptied into the basin. Dissection is greatest in the most northerly fan.

ALLUVIUM

Alluvium occupies the flood plain of the major streams draining the basin. In the valley of the North Fork Malheur River, along the Middle Fork Malheur River west of Drewsey and the center of Otis Basin are the only places it has been concentrated to a depth sufficient for agricultural purposes.

The alluvium consists predominantly of reworked material from the Juntura formation and detritus from the basement complex and Drinkwater basalts. Along the channel of the North Fork Malheur River are cobbles and gravels that have been derived from the plutonic and metamorphic rocks of the region to the north.

A landslide deposit occurs at the north end of Pete Mountain. Large blocks of basalt mixed with tuff form a hummocky topography along the river. At the landslide the river changes from a meandering stream to an actively cutting youthful stream. The landslide is probably a result of the downcutting of the stream, which has over-steepened the valley walls.

STRUCTURAL GEOLOGY

FOLDING

A northwest trending anticline with its axis just east of Drinkwater Pass is a prominent structure in the Juntura Basin. The anticline is asymmetrical, with the west limb dipping more steeply than the east. The anticline forms a highland with the highest elevation of the area near the crest of the fold. The fold is frequently faulted on the east, with the result that the rocks exposed on the eastern limb are older than those at the axis of the anticline.

To the west a shallow syncline very nearly parallels the anticline. Sediments have accumulated in this syncline to a present thickness of approximately 1,000 feet since middle Pliocene time. Prior to this structure, Juntura formation sediments and pyroclastics accumulated in local basins with no apparent orientation. It is difficult to locate the axis of the syncline precisely. The axis passes through Table Mountain, but the dips here are too low and variable to project. The syncline is faulted, which makes the interpretation of dips and strikes difficult since they may reflect the faulting as well as the folding.

The folding of the area began before the deposition of the Drewsey formation and continued until at least late Pliocene.

FAULTING

The topography of the area is in part controlled by a series of northerly and northwesterly trending faults. The faulting occurred at two intervals, once before the deposition of the Juntura formation and later after the deposition of the Drewsey formation.

The earlier faulting can be seen on the eastern edge of the basin, where the basement complex is best exposed. The large fault east of the North Fork Malheur River is a hinge fault with the down-throw side on the west, and the hinge point at the north end of the fault. This fault occurred prior to the deposition of the Juntura formation. This period of faulting which followed the extrusion of the basement basalts was probably a contributing factor in the formation of the Juntura Basin.

Later north trending faults displaced the sediments of the Juntura formation leaving a prominent scarp on the west side of the Agency Valley Reservoir. These later faults which occur throughout the area generally do not have as much displacement as the older faults.

Fault traces are generally in three directions, N50°W, N50°E, and almost due north. These faults are probably related to other basins within the Northern Great Basin where similar fault patterns have been reported.

The faults form small blocks or wedges which act independently of each other. Some of these blocks have had horizontal movement in addition to vertical movement. Movement along the faults in the western Juntura Basin is often less than 100 feet. This type of faulting is easily observed from the aerial photographs of Stinkingwater Mountain. Northwesterly faults are the most common.

Numerous faults can be seen in the sedimentary units along U. S. Highway 20. They are of small displacement and not apparent on the surface. Faults in the sediments are quickly obliterated, and therefore sometimes difficult to locate in the field.

GEOLOGIC HISTORY

The origin of the Juntura Basin as a site of deposition is the result of both intensive erosion and some faulting of the basement complex. The erosion is revealed in a number of ways. At several localities where the basement complex is exposed in section, the basalt which makes the uppermost part is deeply weathered. Considerable relief is apparent in the basin before the deposition of the Juntura formation. This relief was great enough that all members of the Juntura formation are at some point in contact with the basement even though they show no angular difference. Less commonly, some members of the Drewsey are also in contact with the basalts of the basement. Perched remnants of the Drewsey and Juntura formations on the west slope of Pete Mountain indicate that much of this large exposure of the basement complex has always been topographically higher than the succeeding rocks deposited in the basin. The Juntura and Drewsey formations did not completely bury higher elevation topographic features of the basement complex. Lower elevation irregularities were buried by the later sediments and volcanics, but have subsequently been exposed by erosion or later faulting.

The prominent fault east of the North Fork Malheur River apparently was active in the origin of the basin although much of the movement expressed was subsequent to filling of the basin.

Volcanic activity early in the filling of the basin is reflected in the thick ash beds of the lower Juntura formation. In the northern part of the basin, in particular, agglomerates and a basalt indicate that the nature or possibly the site of the activity changed near the end of time represented by the lower member of the Juntura formation. The lack of any large amounts of ash in the succeeding diatomite of the middle-member indicates that volcanic activity was at a minimum during this time. The thick accumulation of diatomite in the central basin indicates that the relatively large bodies of quiet water were present and that drainage had been disturbed in order that these bodies could form. The earlier volcanic activity may account for the formation of dams in existing drainages.

The sediments of the upper member of the Juntura formation indicate that drainage had been re-established either by filling of the local lake basins or by diastrophism. Since this appears as a gradual change in the character of the sediments, the former explanation seems more likely. Some volcanic activity was apparent in the Clarendonian, for ashes make up an important part of these sediments. Some of these, because of their purity, such as the prominent vitric tuff, do not represent redeposition of ashes of earlier volcanic activity.

The major folding began at this time accompanied by erosion and reworking of the Juntura formation sediments. This resulted in a low relief much less than that seen in this area today. As these processes continued they resulted in the accumulation of the Drewsey formation. Some volcanic activity accompanied the early deposition of the Drewsey formation as is evidenced by the welded tuff, agglomerate, and basalt in the lower part of this unit. In the course of deposition of the Drewsey formation, the Black Butte intrusive was emplaced. It is not yet clear whether this intrusive activity was directly related to the various vitric tuffs and the welded tuff or not.

By this time folding ceased. The erosion of higher areas and the deposition of this debris in the synclinal area and other low-relief areas continued somewhat longer. After the deposition of the Drewsey formation, the limbs of the folds were fractured and displaced. The extrusion of the Drinkwater Basalt followed and was in turn followed by local rhyolitic intrusions in the western part of the basin.

The resulting terrain has been eroded since that time. Alluvial fans, terraces, and stream alluvium have resulted from this continuing process.

REFERENCES

BRAMLETTE, M. N. 1946. The Monterey formation of California and the origin of its siliceous rocks. *U. S. Geol. Surv. Prof. Paper* 212.

CHANEY, R. W. 1959. Miocene floras of the Columbia plateau. *Carn. Inst. Wash. Pub.* 617 (1): 1–134.

DOWNS, T. 1956. The Mascall fauna from the Miocene of Oregon. *Univ. Cal. Pub. Geol. Sci.* 31 (5): 199–354.

FENNER, C. N. 1923. The origin and mode of emplacement of the Great Tuff Deposit in the Valley of Ten Thousand Smokes. *Nat. Geog. Soc. Tech. Papers Katmi Ser.* No. 1.

GEORGE, W. D. 1924. The relation of the physical properties of natural glasses to their chemical composition. *Jour. Geol.* 32: 353–372.

GILBERT, C. M. 1938. Welded tuff in eastern California. *Geol. Soc. Amer. Bull.* 49: 1829–1861.

MOORE, B. N. 1937. Nonmetallic mineral resources of eastern Oregon. *U. S. Geol. Sur. Bull.* 875: 1–180.

PIPER, A. M., T. W. ROBINSON, C. F. PARK. 1939. Geology and groundwater resources of the Harney basin, Oregon. *U. S. Geol. Sur. Water Supply Paper* 841.

ROBERTS, R. J. 1956. Flowage structure in central Nevada ignimbrites. *Geol. Soc. Amer. Bull.* 67: 1780.

RUSSELL, I. C. 1903. Preliminary report on artesian basins in southwestern Idaho and southeastern Oregon. *U. S. Geol. Surv. Water Supply Paper* 78: 37–38.

3. MOLLUSKS OF THE BLACK BUTTE LOCAL FAUNA *

D. W. TAYLOR

U. S. Geological Survey

Mollusks of the Black Butte local fauna have been collected at three localities in the upper Juntura formation (see map 2), each representing a similar assemblage of a few aquatic species. Their occurrence is summarized in table 3.

Only a few species are represented in this fauna, and as most of them are either indeterminate or known only from the Black Butte local fauna they cannot contribute nearly as much as the mammals to determining the age of that assemblage.

Viviparus turneri is known from a number of other localities where independent mammalian and stratigraphic evidence suggests a Barstovian or Clarendonian age. The occurrence of *V. turneri* in the Clarendonian Black Butte local fauna thus agrees with previous knowledge of its range.

The environment inferred from the mollusks is a perennial, shallow, fresh-water lake. All of the genera are aquatic, and have a low tolerance for salinity. Six of the nine species are gill-breathers, requiring perennial fresh water for respiration. *Carinifex*, although a pulmonate, is also characteristic of perennial water bodies.

A lacustrine environment is suggested not only by the thin, even bedding of the fine-grained sediments at localities 19116 and 21173, but also by the mollusks. *Carinifex* is the most abundant of the fossil mollusks, and hence its habitat range is of special significance. The known fossil and Recent occurrences of the genus indicate that it is characteristically lacustrine, but may live in perennial streams. *Viviparus* and *Radix* are found in lakes and streams, but the former is today more abundant in lakes. The other genera have a somewhat greater ecologic range, with various species

* Publication authorized by the Director, U. S. Geological Survey.

in various habitats. They do not, however, oppose the inference of a lacustrine habitat.

The depth of water suggested by the fossils is about 15–30 feet. The fauna probably lived within the zone of rooted vegetation, and the condition of the fossils does not suggest any significant transportation.

SYSTEMATIC PALEONTOLOGY

PHYLUM **MOLLUSCA**

CLASS **PELECYPODA**

FAMILY **SPHAERIIDAE**

Genus *Sphaerium* Scopoli, 1777

Sphaerium cf. *S. lavernense* Herrington, 1958

Occurrence: Black Butte local fauna; and unnamed sediments in the Mitchell Butte quadrangle, Malheur Co., Oregon.

Material: Loc. 19116 ($\frac{1}{2}$) ; loc. 19117 (4+ 24/2; all internal molds) ; loc. 21173 (12/2; 3 left valves, 9 right valves).

Remarks: The small amount of well-preserved material from the Black Butte local fauna evidently represents the same species as a much larger series from locality 21175 in the Mitchell Butte quadrangle. The species is closely related to or the same as *S. lavernense* Herrington (Herrington and Taylor, 1958) from the Clarendonian Laverne local fauna of Oklahoma.

Genus *Pisidium* Pfeiffer, 1821

Pisidium cf. *P. clessini* Neumayr, 1875

Occurrence: Locality 19116, Black Butte local fauna.
Material: Loc. 19116 (5/2, all left valves).
Remarks: All but one specimen are crushed and distorted, although the external sculpture is well preserved. The poor preservation makes cleaning and examination of the hinge impracticable, so that identification is dependent primarily on external features of one shell. This valve is 5.0 mm. long and 4.3 mm. wide. It has an oval outline, slightly narrower anteriorly, broken by a convex, protuberant beak which is rounded, not pointed. The shell margins are broadly rounded, without angles. All specimens show strong concentric ribbing. The ribs are strongest and most widely spaced on the beak and upper part of the disk, becoming weaker and more crowded toward the margins. The strong, distant ribs are about one-third to one-fourth as wide as their interspaces, which contain several weak growth lines or low riblets, quite inconspicuous compared to the major series of ribs.

TABLE 3

OCCURRENCE OF MOLLUSKS OF THE BLACK BUTTE FAUNA

	19116	19117	21173
Pelecypoda			
Sphaerium cf. *S. lavernense* Herrington	x	x	x
Pisidium cf. *P. clessini* Neumayr	x		
Pisidium indet.	x		x
Gastropoda			
Fluminicola junturae Taylor, n. sp.	x	x	x
Hydrobiidae indet.			x
Viviparus turneri Hannibal	x	x	x
Radix junturae Taylor, n. sp.	x	x	x
Carinifex shotwelli Taylor, n. sp.	x	x	x
Promenetus indet.			x

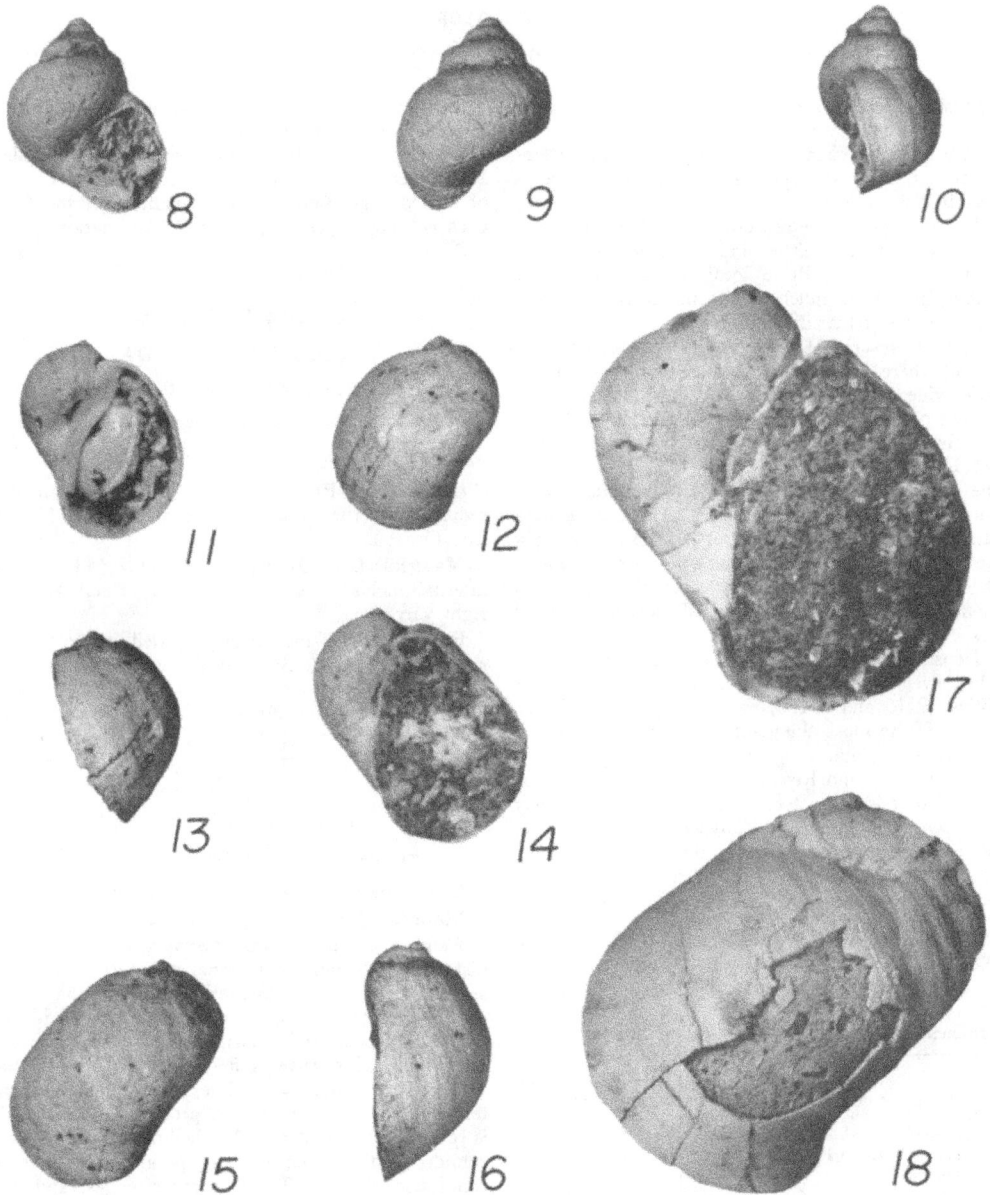

FIGS. 8–18. Gastropods of the Black Butte fauna.

8–10. *Fluminicola juncturae* Taylor, n. sp. Holotype, USNM 563115 ×5. Apertural, dorsal, and lateral views. loc. 21173.

11–13. *Radix juncturae* Taylor, n. sp. ×5, loc. 21173. Holotype, USNM 563107. Apertural, dorsal, and lateral views.

14–16. *Radix juncturae* Taylor, n. sp. ×5, loc. 21173. USNM 563108. Apertural, dorsal, and lateral views.

17–18. *Radix juncturae* Taylor, n. sp. ×5, loc. 21173. USNM 563109. Apertural, and dorsal views.

Sculpture of concentric ribbing is rare in *Pisidium*, but the American *P. ultramontanum* Prime also has concentric ribbing. A series from USGS locality 21175 in Malheur County, Oregon, identified by H. B. Herrington, not only shows the early Pliocene or late Miocene occurrence of *P. ultramontanum* but also helps to indicate the specific distinctness of the fossils from the Black Butte local fauna. These *P. ultramontanum* are about two-thirds as large as *P.* cf. *P. clessini*, of trigonal-oval form, with more prominent, narrower beaks which are more posteriorly situated.

Pisidium clessini has been thoroughly described and illustrated by Woodward (1913) (as *P. astartoides* Sandberger; see Wenz, 1929). The shells from the Black Butte local fauna agree well with his illustrations in external features, and may be conspecific. *P. clessini* has heretofore been known only from the Pliocene and Pleistocene of Europe; however, an examination of the hinges of a series of specimens is necessary to authenticate its American occurrence.

Pisidium indet.

Occurrence: Localities 19116 and 21173, Black Butte local fauna.

Material: Loc. 19116 (8/2; 4 right valves, 4 left valves); loc. 21173 (4/2, all left valves).

CLASS **GASTROPODA**

FAMILY **VIVIPARIDAE**

SUBFAMILY **VIVIPARINAE**

Genus *Viviparus* Montfort, 1810

Viviparus turneri Hannibal, 1912

Campeloma sp.: Turner, 1900a, *Amer. Geologist* **25**: 169. Turner, 1900b, *U. S. Geol. Survey Ann. Rep.* **21** (2): 203.
Vivipara, close to if not *V. couesi*: Spurr, 1905, *U. S. Geol. Survey Prof. Paper* **42**: 67.
Viviparus (Callina) turneri: Hannibal, 1912, *Malacol. Soc. London Proc.* **10**: 193, 202, pl. 8, fig. 31.
Viviparus turneri Hannibal: Buwalda, 1914, *Calif. Univ. Dept. Geol. Bull.* **8**: 351. Merriam, 1916, *ibid.* **9**: 165. Henshaw, 1942, *Carnegie Inst. Wash. Publ.* **530**: 84, 85, 89.
?*Viviparus,* sp.: Hanna, 1922, *Oregon Univ. Publ.* **1** (12): 13.
Viviparus or *Campeloma*: Stearns, 1955, *Geol. Soc. Amer. Bull.* **66**: 1681.

Holotype: UCMP 12216. "Near coal-mine," Silver Peak Range, Esmeralda Co., Nevada. H. W. Turner coll.

Diagnosis: Shell large (40–45 mm. long), narrowly conic, with six strongly convex whorls separated by an incised suture, an ovate aperture, and imperforate base. A well-marked peripheral angulation becomes obsolete at about the third whorl, after which the contour of the shell is smoothly and regularly rounded. Preserved sculpture consists of irregular, slightly retractive growth lines.

Nearly all of the adult specimens are decollated, crushed, or represented by internal molds, and hence reliable measurements cannot be given for most. One large adult specimen from locality 6540 measures as follows (mm.): length 50; width 33½; length of aperture 25½; width of aperture 22; 6 whorls. In size and proportions this specimen and others from the same locality are similar to less well-preserved specimens from the other localities.

Occurrence: *Viviparus turneri* is known from the Barstovian and Clarendonian, in the western Great Basin, southeastern Oregon, and adjacent Idaho. Independent dates based on fossil mammals are available for the Black Butte local fauna, the Tonopah local fauna, and Cedar Mountain local fauna. The following list summarizes the known occurrences of *V. turneri*, and indicates the source of independent age assignments. The material examined is in the collections of the University of California Museum of Paleontology (UC) and U. S. Geological Survey (USGS).

A. Barstovian, Tonopah local fauna, San Antonio Mountains, Nye Co., Nevada (Henshaw, 1942).

USGS Cenozoic loc. 3706. 8 miles north of Tonopah and 1 mile west of the mining camp of Ray. J. E. Spurr, 1902. Henshaw (1942: 87) recorded a more precise location for Spurr's collection and gave a measured section.

USGS Cenozoic loc. 10169. 3.4 miles N. 30° W. from Monument 206. H. G. Ferguson, 1922.

USGS Cenozoic loc. 10298. 8½ miles north of Tonopah. H. G. Ferguson, 1922.

B. Clarendonian, Black Butte local fauna, Malheur Co., Oregon. Specific localities and mammalian evidence for age are given elsewhere in this volume.

C. Clarendonian, Cedar Mountain local fauna, Tonopah, quad., Nye Co., Nevada (Buwalda, 1914; Merriam, 1916). Mollusks were collected at a number of localities by C. L. Baker and J. P. Buwalda in 1912, and recorded by Buwalda without specific locality data. The locality descriptions are based on both published information (Merriam, 1916) and on the University of California Museum of Paleontology catalogues.

UC 1971. At 6250 feet elev., 1 mile north of road intersection which intersection is 2 miles SE of Bell Spring.

UC 1984. About 3 miles below (NW) of Stewart Spring. At the base of the "s" of the word Wash, of Finger Rock Wash on U.S.G.S. Tonopah sheet. In the next layer above the contorted pumice layer, which last overlies material which looks like shoreline deposits.

UC 2022. One-fourth mile S and SE of 2021, and very nearly the same horizon. UC 2022 is around flanks of conical hill at E end of small ridge.

UC 2021: On a little white hill in middle of canyon (hill is 50–60 feet high), conical in shape, white all around. Location by Brunton: S. 56° E. of little

Pilot Peak; N. 28° E. of dome which is enclosed by 6700 contour line (dome is roughly 2 miles east of Bell Spring).

D. Localities without independent stratigraphic or paleontologic dates.

USGS Cenozoic loc. 20252. Washington Co., Idaho. NE ¼ sec. 27, T. 11 N., R. 7 W. Payette formation. Coal Mine Gulch, north bank of Snake River, H. T. Stearns coll. The single specimen is badly crushed, but preserves the outline and shows the size and the lack of prominent sculpture. The specimen was recorded by Stearns (1955) as suggesting possible Cretaceous age for the fossil-bearing unit.

USGS Cenozoic loc. 21175. Malheur Co., Oregon. Mitchell Butte quad. (1906) 1:125000. NE¼ sec. 21, T. 23S., R. 43 E., unsurveyed. 3,400 feet elev. (aneroid). R. E. Corcoran, 1955.

USGS Cenozoic loc. 6540. Humboldt Co., Nevada. From foothills just south of Sulphur station, Western Pacific R.R. R. Forrester, 1910.

USGS Cenozoic loc. 8068. Mineral Co., Nevada. Lake beds near Ourco, east of Mina. Fred Siebert coll.

USGS Cenozoic loc. 2021. Esmeralda Co., Nevada. 10 miles south of Columbus. C. D. Washburn coll.

UC 3879. Esmeralda Co., Nevada. Silver Peak Range, 1 mile SE of coal mine. "Knapp's locality." H. W. Turner coll.

UC 3876. Silver Peak Range, about 1½ miles SE of coal mine. J. D. Reed, 1899.

UC 3881. Silver Peak Range, 135° from coal mine. About 300 feet back from mine. J. D. Reed, 1899. Type locality of *Viviparus turneri.*

UC V2803. Esmeralda Co., Nevada. Tonopah quad. T. 2 N., R. 37 E. 4 miles WNW Blair Jn. Fine grained buff sandstone which forms a small strike ridge (Dip 25° strike N. 55° W.). Fish and shells. Vanderhoof, 1928.

Material: Loc. 19116 (42); 19117 (26); 21173 (20).

FAMILY **HYDROBIIDAE**
SUBFAMILY **HYDROBIINAE**

Genus *Fluminicola* Stimpson, 1865

Fluminicola junturae Taylor, n. sp.

Holotype: USNM 563115. Locality 21173, Black Butte local fauna Malheur Co., Oregon.

Diagnosis: Shell conic, with acute spire and rounded base; whorls 4–4¼, convex, separated by an impressed suture; aperture no more than half of total shell length. Named for the Juntura Basin in which it occurs.

Description: The shell is conic, with an acute spire and varies from narrowly conic to globosely so; the base is rounded, with a narrow umbilical chink. The 4–4¼ moderately convex, smoothly rounded whorls are

separated by an impressed suture. The aperture is oval and simple; broadly rounded below, obtusely subangular above, and no more than half of the total shell length. Sculpture consists only of fine growth lines.

Measurements (in mm.) of the holotype, the only complete, undistorted adult specimen available, are as follows: length, 6.6; width 5.3; length of aperture, 3.3; width of aperture, 2.9; 4 whorls.

Occurrence: Known only from the Black Butte local fauna, Malheur Co., Oregon.

Material: Loc 19116 (5); 19117 (42); 21173 (43).

HYDROBIIDAE indet.

Material: Loc. 21173 (5).

Remarks: A small hydrobiid is of uncertain generic position because of the present lack of knowledge of western American representatives of the group. It may be related to *"Paludestrina" stearnsiana* Pilsbry. An undistorted, complete, and probably adult specimen is a small, smooth, narrowly conical, coiled tube with blunt apex, measuring (mm.) as follows: length, 1.6; width, 1.25; length of aperture, 0.7; width of aperture, 0.6; 2¾ whorls.

FAMILY **LYMNAEIDAE**

Genus *Radix* Montfort, 1810

Radix junturae Taylor, n. sp.

Holotype: USNM 563107. Locality 21173, Black Butte local fauna, Malheur Co., Oregon.

Diagnosis: Shell ovate, small for the genus, thick, with a large, effuse body whorl in length 75–90 per cent of the total shell length; whorls three, smoothly rounded; suture sharply impressed.

The species is named for the Juntura Basin in which it occurs.

Description: The shell is ovate to globosely ovate, small for the genus, and relatively thick. The three smoothly rounded whorls expand rapidly so that the body whorl is very large and effuse; in length they are 75–90 per cent of the total shell length. The aperture is oval to almost semicircular, and simple. The parietal lip is appressed to the preceding whorl, and variably reflected so as to leave a narrow umbilical slit or none. The columella is weakly twisted; the columellar wall of

---

19–22. *Carinifex shotwelli* Taylor, n. sp. ×5 loc. 21173. Holotype, USNM 563110. Apertural, lateral, apical (21) and basal (22) views.

23–26. *Carinifex shotwelli* Taylor n. sp. ×5 loc. 21173 USNM 563111. Apertural, lateral, apical (25), and basal (26) views.

27–30. *Carinifex shotwelli* Taylor. n. sp. ×5 loc. 21173 USNM 563112. Apertural, lateral, apical (29), and basal (30) views.

31–34. *Carinifex shotwelli* Taylor, n. sp. ×5 loc. 21173. USNM 563113. Apertural, lateral, apical (33), and basal (34) views.

35–36. *Carinifex shotwelli* Taylor, n. sp. ×5 loc. 21173. USNM 563114. Apertural and apical views.

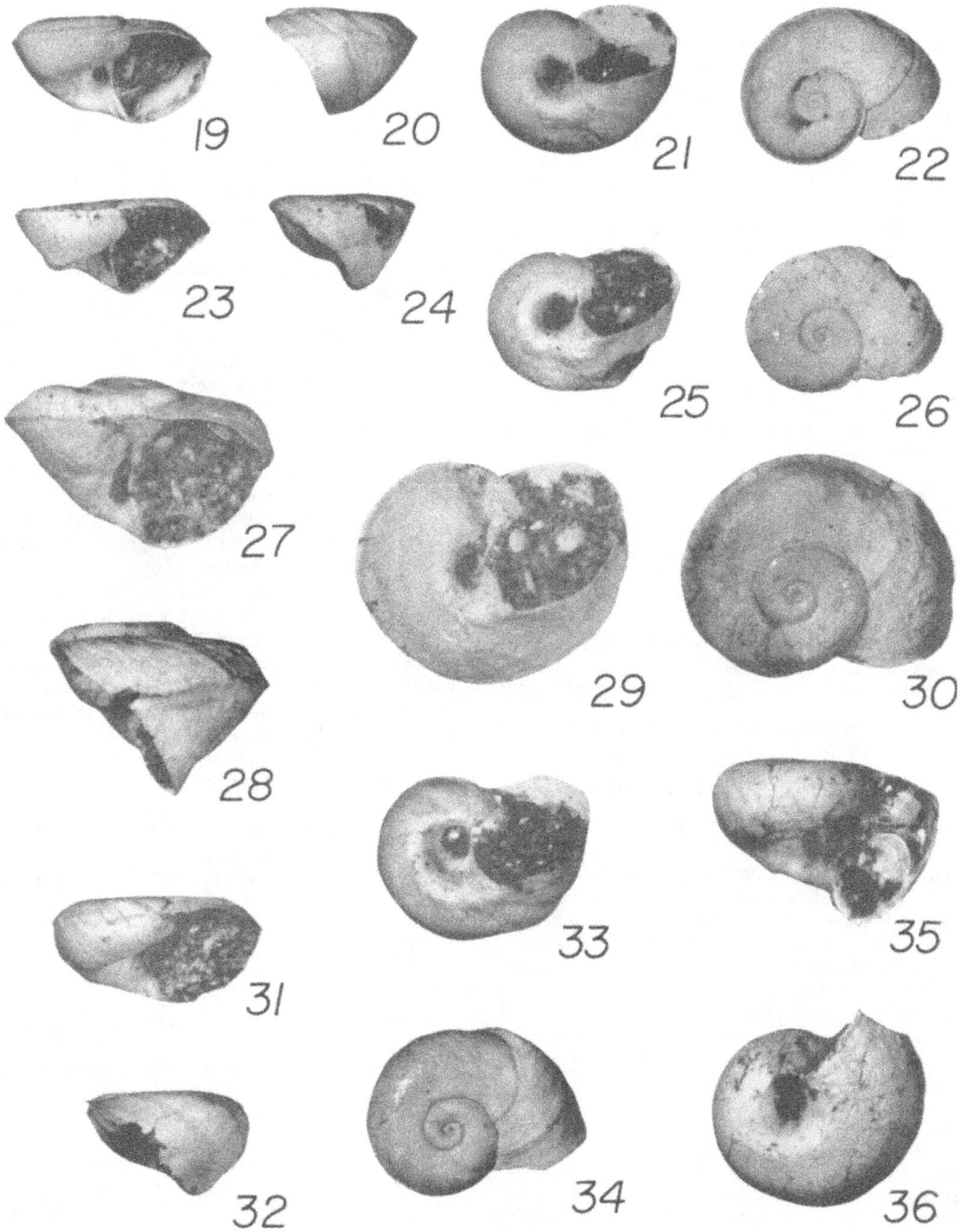

FIGS. 19–36.   Gastropods of the Black Butte fauna.

the aperture may show a weak plication, or none at all. The short, inconspicuous spire consists of about two whorls separated by a finely incised suture. The spire whorls are strongly convex, so that short-spired shells have a smoothly rounded spire scarcely breaking the contour of the body whorl. The only sculpture consists of irregular axial growth lines, fine to moderately coarse, slightly retractive toward the base.

Reliable measurements are difficult to obtain because most of the shells are crushed. The following dimensions (in mm.) are taken from the holotype, an immature but undistorted shell; and are estimated for two larger specimens (figured) which are crushed.

TABLE 4

MEASUREMENTS OF *Radix junturae* TAYLOR

	Length	Width	Length aperture	Width aperture	No. whorls
Holotype	6.6	5.9	6.0	4.0	2¼
USNM 563108	8.2	7.1	7.2	5.0	2¼
USNM 563109	14.3	12.5	11.5	8.5	3

Variation: The chief variants are rate of whorl growth and inclination of whorls. Thus the aperture varies from oval to almost semicircular, and the spire may be one-quarter or almost none of the total shell length.

Occurrence: Known only from the Black Butte local fauna, Malheur Co., Oregon.

Material: Loc. 19116 (2) ; 19117 (13) ; 21173 (75).

FAMILY **PLANORBIDAE**

SUBFAMILY **HELISOMATINAE**

Genus *Carinifex* Binney, 1863

Carinifex shotwelli Taylor, n. sp.

Holotype: USNM 563110. Locality 21173, Black Butte local fauna, Malheur Co., Oregon.

Diagnosis: Shell small for the genus, hemispherical; right side plane; left side strongly convex with a narrow, deep spire-pit; periphery strongly carinate to rounded; whorls 3–3½.

Named for Dr. J. Arnold Shotwell, Museum of Natural History, University of Oregon.

Description: The shell is small for the genus, roughly hemispherical in shape. The right side is plane or slightly concave; the left side strongly convex, with a narrow, deep spire-pit whose width is contained 3–3½ times in the shell diameter. The left side of the body whorl may be evenly convex adjacent to the sharp angle bordering the spire pit, or it may be slightly constricted around this angle, so that in cross section the base of the whorl appears produced. The periphery of the shell may be broadly rounded, but usually it bears a sharp carina produced as an elevated ridge rather than simple angulation. This peripheral carina

TABLE 5

MEASUREMENTS OF *Carinifex shotwelli* TAYLOR

Cat. no.	Height	Width	Height aperture	Width aperture	No. whorls
USNM 563114	6.0	8.3	—	3.6	3½
USNM 563112	6.4	9.7	5.3	4.6	3¼
USNM 563113	4.0	7.9	3.6	4.0	3
USNM 563110	4.0	7.0	3.1	3.3	3
USNM 563111	3.7	6.8	3.3	3.3	3

is closer to the right side than to the median plane, but its position varies with the convexity of the right side. The aperture is oval-quadrate to trapezoidal; sometimes the lower part of the outer lip is constricted and the base produced. The whorls number 3–3½, and are separated by a weakly to moderately impressed suture. Axial sculpture consists only of fine, irregular growth lines, retractive toward the base. Spiral sculpture consists of the peripheral carina and of numerous fine, irregular, impressed lines which cut the growth lines. On worn specimens these are not preserved, but when observable they give the shell surface a glossy texture.

Measurements (in mm.) are given for the five figured specimens in table 5.

Variation: The most conspicuous variation is in the shape of the whorls and in the strength of the carina. Both the measurements and illustrations show the range of variation in whorl shape. The whorls may be constricted, with almost flat surfaces and sharp angles, or may be swollen and more rounded. The carina may be almost at the extreme right side of the shell, or displaced slightly toward the median line. Most shells have a well-developed carina, but it is weak to absent on some.

Occurrence: Known only from the Black Butte local fauna, Malheur Co., Oregon.

Material: Loc. 19116 (7) ; 19117 (23) ; 21173 (500).

SUBFAMILY **PLANORBULINAE**

Genus *Promenetus* Baker, 1935

Promenetus indet.

Occurrence: Black Butte local fauna, loc. 21173.

Material: Loc. 21173 (3).

Remarks: The material is specifically indeterminate, but similar in shape to *Promenetus exacuous* (Say) and *P. kansasensis* (Baker).

REFERENCES

BUWALDA, J. P. 1914. Tertiary mammal beds of Stewart and Ione Valleys in west-central Nevada. *Calif. Univ. Dept. Geol. Bull.* **8**: 335–363.

HANNA, G. D. 1922. Fossil freshwater mollusks from Oregon contained in the Condon Museum of the University of Oregon. *Oreg. Univ. Pub.* **1**: (12).

HANNIBAL, HAROLD. 1912. A synopsis of the recent and Tertiary freshwater Mollusca of the Californian province, based upon an ontogenetic classification. *Malacol. Soc. London Proc.* **10**: 112–211.

HENSHAW, P. C. 1942. A Tertiary mammalian fauna from the San Antonio Mountains near Tonopah, Nevada. *Carn. Inst. Wash. Pub.* **530**: 77–168.

HERRINGTON, H. B., D. W. TAYLOR. 1958. Pliocene and Pleistocene Sphaeriidae (Pelecypoda) from the central United States. *Mich. Univ. Mus. Zool. Occasional Papers* **596**.

MERRIAM, J. C. 1916. Tertiary vertebrate fauna from the Cedar Mountain region of western Nevada. *Calif. Univ. Dept. Geol. Bull.* **9**: 161–198. Reprint, 1938. *Carn. Inst. Wash. Pub.* **500 3**: 1362–1402.

SPURR, J. E. 1905. Geology of the Tonopah mining district, Nevada. *U. S. Geol. Survey Prof. Paper* **42**.

STEARNS, H. T. 1955. Discovery of Cretaceous (?) sediments in southwestern Idaho (abstract). *Geol. Soc. Amer. Bull.* **66**: 1681–1682.

TURNER, H. W. 1900a. The Esmeralda formation. *Amer. Geol.* **25**: 168–170.

—— 1900b. The Esmeralda formation, a fresh-water lake deposit, by H. W. Turner, with a description of the fossil plants by F. H. Knowlton, and of a fossil fish by F. A. Lucas. *U. S. Geol. Survey Ann. Rept.* **21** (2): 191–226.

WENZ, W. 1929. Zur Geschichte und Verbreitung des Pisidium clessini Neumayr-astartoides Sandberger. *Archiv fur Molluskenkunde* **61**: 185–189.

WOODWARD, B. B. 1913. Catalogue of the British species of Pisidium (recent & fossil) in the collections of the British Museum (Natural History), with notes on those of western Europe. London (British Museum).

4. MAMMALIAN FAUNA OF THE UPPER JUNTURA FORMATION, THE BLACK BUTTE LOCAL FAUNA

J. ARNOLD SHOTWELL and **DONALD E. RUSSELL**

Museum of Natural History, University of Oregon   Muséum National d'Histoire Naturelle, Paris

OCCURRENCE

Black Butte is a prominent peak six and one-half miles southwest of Juntura, Oregon, in Malheur County. The various localities of the Black Butte local fauna, are found in a series of hills extending from one to three and one-half miles west of Juntura. They are bounded on the north by U. S. Highway 20 and on the south by an old county road which joins the Juntura-Riverside road about one mile south of Juntura at the eastern end. Part of the course of this road follows Black Butte Creek (see map 2).

Russell (1956) working with the materials collected the first two years in the area, described the fauna, as it was then known, in a master's thesis. Subsequently a great deal more material was collected, much of which is more complete than the earlier specimens. Because of this, some determinations of taxonomic affinity will differ from those originally seen in Russell's thesis. However, some important parts of the fauna are described here essentially as they were in the earlier report. Mr. Russell's work then has materially aided in the description of the fauna and for this reason he appears as a junior author. At the time of this writing Mr. Russell is in France so that communication between us concerning affinities of new materials or more complete data is not practical. I am sure, if he could see the additional evidence, that he would concur in the corrections of the preliminary report he authored.

Nearly thirty different localities were established in the Juntura formation. They are indicated on the accompanying map (map 3). Several of these were developed into quarry sites. This chapter will deal primarily with the systematics and age of the fauna. The paleoecology of the occurrences has been described in an earlier chapter.

COMPOSITE LIST OF THE BLACK BUTTE LOCAL FAUNA

Order Insectivora
  Superfamily Soricoidea
    Family Soricidae
      ?Hesperosorex sp.
    Family Talpidae
      Scapanus sp.
Order Lagomorpha
  Family Leporidae
    Hypolagus sp.
Order Rodentia
  Suborder Sciurimorpha

Family Aplodontidae
  Tardontia sp.
Family Mylagaulidae
  Mylagaulus sp.
  Epigaulus minor Hibbard and Phillis (1945)
Family Sciuridae
  Citellus junturensis n. sp.
  Citellus sp.
Family Heteromyidae
  Cupidinomys sp.
Family Castoridae
  ?Hystricops sp.
  Eucastor malheurensis n. sp.
Suborder Myomorpha
  Family Cricetidae
    Peromyscus cf. *dentalis* Hall (1930a)
Order Carnivora
  Family Canidae
    Osteoborus sp.
    Aelurodon sp.
    Vulpes sp.
  Family Mustelidae
    Martes sp.
    Pliotaxidea sp.
    Eomelivora sp.
    Sthenictis junturensis n. sp.
  Family Felidae
    Pseudaeluras sp.
Order Probscidea
  Family Gomphotheriidae
    Platybelodon (*Torynobelodon*) cf. *barnumbrowni* Barbour (1929)
  Family Mammutidae
    Mammut (*Pliomastodon*) *furlongi* n. sp.
    Mammut (*?Miomastodon*) sp.
Order Perissodactyla
  Suborder Hippomorpha
    Family Equidae
      Hipparion condoni Merriam (1915)
  Suborder Ceratomorpha
    Family Tapiridae
      Genus indet.
    Family Rhinocerotidae
      Aphelops sp.
Order Artiodactyla
  Suborder Suiformes
    Family Tayassuidae
      Prosthennops sp.

Family Merycoidontontidae
  Ustatochoerus sp.
Suborder Tylopoda
  Family Camelidae
    Procamelus cf. *grandis* Gregory (1942)
    Megatylopus cf. *gigas* Matthew and Cook (1909)

AGE AND CORRELATION

The association of *Eucastor, Platybelodon, Ustato-choerus, Aelurodon, Procamelus,* and *Hipparion* in the fauna indicate a Clarendonian Age. The advanced stage of development of *Eucastor malheurensis* and the lack of *Hypohippus* suggest that this fauna represents the latter Clarendonian. Comparison with the Black Hawk Ranch fauna of northern California indicates a similar age. This may be in part due to similar local habitat situations since no species are common to both faunas. The Black Hawk Ranch fauna is the type fauna of the Montediablan stage of Savage (1955) which is latter Clarendonian. The Black Hawk Ranch as well as other Montediablan faunas are situated in areas which are very probably distinct biotic provinces from those of the Northern Great Basin in the Clarendonian. Biotic provinces are continuous areas of relatively uniform climate recognized by one or more ecologic associations which differ from adjacent provinces. These provinces are centers of taxonomic differentiation. The stage as used by Savage is "—the stratal occurrence and representation of phyletic change, a peculiar and particular interval in the evolution of land mammals." As stage is a time term and is not necessarily restricted by geography, presumably a particular stage in the evolution of land mammals might be found most any place. However, it is necessary to characterize a stage by species, and thus it becomes as provincial as species. Stages as applied then are limited to biotic provinces. We cannot refer the Black Butte fauna to the Montediablan stage but can correlate it with this stage, recognizing the similarity in stage of evolution of the mammals involved. Similarly we can correlate the Black Butte fauna with the Ricardo and Big Springs Canyon faunas.

The Brady Pocket and Nightingale road faunas of the Truckee formation (Macdonald, 1956) are very similar to the Black Butte local fauna. All of the forms in the faunal list of these Nevada localities are found at Black Butte sites with the exception of the camel *Aepycamelus.* The difference in generic assignment of the horses is one of judgment and not real. The Nevada sites include thirteen genera of mammals whereas the Black Butte fauna is represented by thirty genera of mammals in a much larger collection. The Brady Pocket fauna is particularly noted here because of its close resemblance to the Black Butte fauna. It is the closest known Clarendonian mammal locality in the Northern Great Basin.

RELATIVE STRATIGRAPHIC POSITIONS OF LOCALITIES WITHIN THE JUNTURA FORMATION

The Black Butte local fauna was collected from the upper bedded portion of the Juntura formation. The individual beds, although in some cases covering large portions of the area from which the specimens were collected, are for the most part either impossible to recognize from one exposure to another or are local lenses. The frequent recurrence of similar beds throughout the upper part of the section adds to the difficulty of maintaining stratigraphic controls. Fortunately a persistent pumice band about fifteen feet thick occurs throughout the major part of the area of the localities. This bed, although of loose glass shards, forms a prominent ledge which may be easily traced. Similar pumice bands, usually thinner, occur lower in the section; however there is little doubt which bed is present in a given exposure due to the continuous nature of this marker bed. Presumably this pumice, since it contains little or no impurities, may be considered as a time marker in the section at least for the purposes of determining the relative position of various localities.

All but one of the established localities of the Black Butte local fauna are below this marker pumice. The one above is immediately above the bed. Many of the localities are found just below the pumice marker. These often occur in cross-bedded sand lenses within a massive brown to buff tuff. The major locality, quarry 3 (UO loc. 2448), is found in such a position. A number of localities were established at a level about fifty feet lower in the section. Quarry 11 (UO loc. 2337) apparently is the lowest locality stratigraphically. Figure 7, chapter 2, indicates the relative position of the localities in the generalized section.

Not all localities consisted of the recovery of materials in place. Float materials were given locality numbers separate from specimens in place. They cover a somewhat broader stratigraphic section, usually a portion of an exposure. It was usually possible to determine approximately what portion of the exposure was producing the bones collected but not always within less than twenty feet vertically. The position of quarries 1 and 2 (UO locs. 2328 and 2332) relative to the pumice bed is difficult to determine since the bed is not present in the immediate vicinity of these localities. They are apparently at a similar stratigraphic level to the pumice bed, probably immediately below it.

SYSTEMATICS

Order INSECTIVORA

Family SORICIDAE

?Hesperosorex sp.

A large shrew is present in the fauna from locality 2337. A fragment of a jaw with the M1–M2 in it indicates the presence of this form. The lack of a deep valley between the entoconid and metaconid, a heavy external cingula and the size are very much like these characteristics in *Hesperosorex* Hibbard (1957). The ascending ramus is missing, so it is not possible to compare further with *Hesperosorex lovei*. The internal cingula is more poorly developed and the entoconid is farther from the metaconid than in *H. lovei*. The antero-posterior length of M1–M2 is 3.2 mm. See figure 37.

FAMILY **TALPIDAE**

Scapanus sp.

Several humeri fragments of a mole are in the collection from locality 2337. These compare favorably with the humerus of *Scapanus*.

ORDER **LAGOMORPHA**

FAMILY **LEPORIDAE**

Hypolagus sp.

The rabbit *Hypolagus* is represented by a number of loose dental elements, metapodials, calcania, astragali, and fragments of pelvi, tibia, and a humerus. In the teeth the reentrant angle of the upper molars has crenulated borders. There is often a small isolated enamel lake just beyond the reentrant angle in these teeth. An upper second premolar has two prominent grooves running vertically along the anterior face of the tooth. One of these grooves appears as a small reentrant on the occlusal surface. All of these characteristics appear on a number of species of *Hypolagus*.

ORDER **RODENTIA**

FAMILY **APLODONTIDAE**

Tardontia sp.

A fragment of a rooted upper premolar F-10978, UO 2349, indicates the presence of an aplodontid in the fauna. The failure of the enamel on one side of the tooth, well-developed roots, the fossette pattern, and development of the styles are much like *T. nevadans*. However, the tooth is much higher crowned than the homologous tooth in that species. *T. occidentalis* is a higher-crowned species but no upper teeth have yet been assigned to this species. The Black Butte specimen probably represents this species and certainly represents *Tardontia*.

FAMILY **MYLAGAULIDAE**

Mylagaulus sp.

Mylagaulids are common rodents in the Juntura fauna. The material has been described in an earlier work on this group (Shotwell, 1958).

Epigaulus minor Hibbard and Phillis (1945)

Two upper fourth premolars apparently left and right of a single individual were collected at loc. UO 2343. These teeth are described and illustrated in Shotwell (1958).

FAMILY **SCIURIDAE**

Citellus junturensis n. sp.

Type: F-5871—Right lower jaw with incisor, P4-M3, ascending ramus missing.

Referred specimens: F-5763 UO loc. 2341 fig. 38, right lower jaw with incisor, P4-M3 ascending ramus missing.

F-12070—left maxillary fragment with P3-M1, UO loc. 2334, fig. 39.

Type locality: Loc. UO 2338 Juntura formation.

Diagnosis: Small citellid. Protoconid and parametaconid of lower P4 in contact. P4 less molarlike than in other ground squirrels. Hypoflexid *U*-shaped at occlusal surface. Trigonid basin closed on M1 and M2, open on M3. Metalophid does not reach basin floor on M3. No metalophid or protolophid present on P4. Basin floors smooth, not rugose. Size of larger living species of *Ammospermophilus*. Teeth small for size of jaw. Upper P4 not molariform. P3 single cusped and more than one-half antero-posterior diameter of P4.

Description of lower dentition: P4 very small. Protoconid and parametaconid in contact with each other. No indication of protolophid or metalophid. Trigonid only slightly higher than hypoconid, posterolophid smooth, entoconid not recognizable.

M1 larger than P4. Protoconid and parametaconid well separated. Protolophid and metalophid both complete. Metalophid connected to parametaconid. Small but deep trigonid basin. Protoconulid well developed. Entoconid more prominent than in P4 but still only poorly developed. A very small hypoconulid is present in one specimen. Very small mesostylid, may not always be present. Trigonid cusps about same height as hypoconid.

M2 similar in size and shape to M1. Trigonid basin is larger than on M1. Trigonid basin is complete. Prominent protoconulid. Very small mesostylid, may not always be seen. Metalophid not as strongly connected to parametaconid as on M1. On one specimen the connection is weak.

M3 longer than other molars but same width. Trigonid basin is open. However, the metalophid does not extend to floor of talonid basin. Hypoconid much heavier than in other molars.

---

37. *Hesperosorex* sp. UOMNH F-20383, loc. 2337. Labial side ×10 (37*a*), lingual side ×10 (37*b*), and occlusal view of dentition ×20 (37*c*).
38. *Citellus junturensis* Shotwell and Russell, n. sp. Holotype UOMNH F-5871, loc. 2338. Right lower jaw ×5 (38*a*). Occlusal view of dentition ×10 (38*b*).
39. *Citellus junturensis* Shotwell and Russell, n. sp. Assigned specimen UOMNH F-12070, loc. 2334. Occlusal view upper dentition P3-M1, ×15.
40. *Citellus (O)* sp. Upper M1, UOMNH F-12071, loc. 2337. Occlusal view ×15.
41. *Cupidinomys* sp. Lower P4. UOMNH F-18236, loc. 2337. Side view ×20 (41*a*). Occlusal view ×20 (41*b*).

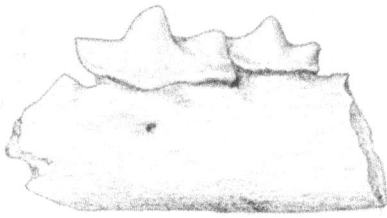

37a

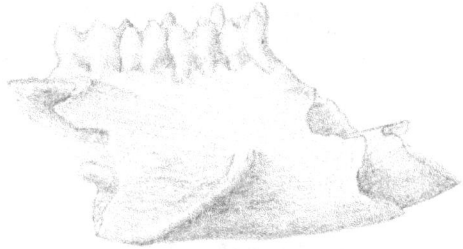

38a

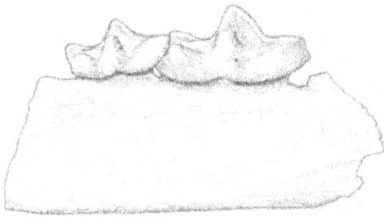

37b

38b

37c

39

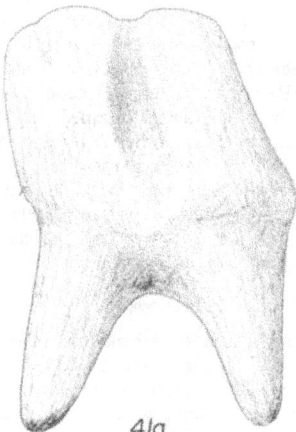

41a

41b

40

FIGS. 37–41. Insectivora and Rodentia of the Black Butte fauna.

Posterior lingual border smooth in molars. Entoconid not prominent.

Description of upper dentition: P3, single cusped peglike tooth.

P4, parastyle very small, only a slight rugosity on anterior cingulum. Does not affect the occlusal outline of the tooth. Anterior cingulum poorly developed with a very narrow and shallow valley, not lower than the posterior cingulum. Paracone prominent and connected to protocone by straight protoloph. Protoconule poorly developed or absent and not noticeable at stage of wear of specimens. Metacone about same height as paracone. Metaconule small. The metaloph is high between metacone and metaconule and low between metaconule and protocone. Connection between metaconule and protocone very low. Protocone is narrow. The small posterior and anterior cingula do not add to its size. The posterior cingulum is about the same size as the anterior. The posterior valley is broader lingually than labially and very flat. No mesostyle. The paracone and metacone are close together giving the tooth an ovoid occlusal outline.

M1, much larger than P4. Parastyle prominent forming a significant part of the occlusal outline of the tooth. Parastyle much lower than paracone. Anterior cingulum well developed with a valley similar to the central valley in size. The anterior cingulum turns abruptly into the protocone and has a small ventro-dorsal groove at its connection with the protocone. The paracone is connected to the protocone by a straight protoloph. No protoconule is evident. The metacone is connected to the protocone by a straight metaloph. There is a metaconule. Unlike the P4, the central valley does not connect with the posterior valley. The posterior cingulum connects smoothly with the protocone and contributes to the size of the protocone. There is no mesostyle present. The occlusal shape of the tooth is that of a U.

The posterior border of the zygomatic arch is between the P4 and M1. The masseteric tubercule is large and immediately ventral to the infraorbital foramen. The infraorbital foramen is not complete in this specimen. It is rounded ventrally and dorso-ventrally elongated. The median dorsal border is missing.

Discussion: The three specimens discussed here came from three separate localities in the Juntura formation. The two lower jaws are nearly identical. The maxillary fragment is assigned to the same species because it represents a ground squirrel of the same size. It also indicates a squirrel in a similar stage of evolutionary development. All the specimens represent a squirrel with small teeth for its total size. Although the upper and lower elements are not found in association, they appear to belong to the same species of squirrel. This new squirrel is smaller than most fossil citellids known. *Citellus ridgewayi* (Gazin, 1933) from Skull Springs, Oregon, is about the same size. The new species differs from *C. ridgewayi* and *C.*

quatalensis in the absence of mesostyles on the upper teeth. The upper P3 is large in relation to the other cheek teeth as in *C. ridgewayi*. The lower dentition of *C. ridgewayi* is not known so that these comparisons cannot be made. The appressed parametaconid and protoconid of the lower P4 and the small size suggest chipmunk characteristics such as those of *Ammospermophilus*. However, the similarity may be misleading since chipmunks can be characterized as retaining many primitive characters of the dentition. This suggests phylogenetic relationship which in fact may be more parallel than direct.

C. junturensis is certainly closely related to *C. ridgewayi* and possibly is derived from the older species. It differs in a number of important characteristics. These include lack of mesostyles, U-shaped trigonid on M1, posterior cingulum broadly connected to protocone, paracone, and metacone close together in both P4 and M1. Protoloph and metaloph nearly parallel.

TABLE 6

MEASUREMENTS OF *Citellus junturensis*
(IN MILLIMETERS)

Lower jaws		I	P4	M1	M2	M3	P4-M3
F-5763	AP	1.8	1.3	1.5	1.8	2.2	6.4
	T	0.9	1.3	1.7	1.9	1.9	
F-5871	AP	2.0	1.3	1.6	1.7	2.2	6.6
	T	0.9	1.3	1.7	2.0	2.0	
Upper jaws		P3	P4	M1			
F-12070	AP	0.7	1.2	1.6			
	T	0.7	1.7	2.1			

Depth of the mandible below M1							
F-5763	5.4						
F-5871	5.3						

Protospermophilus of the late Miocene, a likely ancestor of modern citellids, differs from later citellids in its squared entoconid borders giving it the appearance of tree squirrels in this respect. It also possesses in the lower P4 appressed parametaconid—protoconid. This is an early character retained in *Ammospermophilus*. However, in other Pliocene and recent citellids these two cusps are separated and in some cases widely separated. It is very possible, judging from the material known, that the present diversity of citellids dates to the late Miocene. This has previously been indicated by Bryant (1945: fig. 48).

Citellus (Otospermophilus) sp.

An upper right first or second molar from locality UO 2337, F-12071, is referred to *Citellus (Otospermophilus)* sp. The anterior and posterior cingula form a U-shaped outline with the protocone. The hypocone cannot be distinguished from the protocone. A small mesostyle is present. The anterior cingulum is lower than the posterior cingulum. The protoloph is smooth apparently lacking a protoconule. The paracone is the highest cusp. The metacone is separated by a valley from the metastylid. The metaloph terminates in a well-developed metaconule. The metaloph does not reach the protocone thus allowing the central and posterior valleys to be widely connected. The anterior and central valleys are about the same size. The posterior valley is smaller labially. The parastyle is well developed but separated from the paracone. Antero-posterior diameter 2.0 mm., width 2.8 mm. See fig. 40.

FAMILY HETEROMYIDAE
Cupidinomys sp.

A lower P4 (F-18236) from locality 2337 is referable to *Cupidinomys*. The anterior loph of the tooth is about twice the antero-posterior diameter of the posterior loph. There is a deep notch in the anterior face of the tooth partially closed at the base by a small cusp. The anterior edges of the notch are bulbous. Reference of this specimen to any known species is difficult. It is only little worn and shows many characteristics not seen in more worn dentitions of known species. Ap-1.4 mm., T 1.0 mm. See fig. 41.

FAMILY CASTORIDAE
SUBFAMILY CASTOROIDINAE
?Hystricops sp.

Two large upper fourth premolars from localities UO 2340 and UO 2332 represent a large beaver in the fauna. One of the teeth is well worn whereas the other is virtually unworn. These teeth represent the same species of rodent as that described by Macdonald from "Brady Pocket" (1956). Macdonald refers his material to *Hystricops*. However, the type material of *Hystricops* is well worn. Well-worn castorid teeth lack many of the characters useful in accurate assignment of specimens to even the proper genus.

The unworn upper fourth premolar in the Black Butte collection is excellent material for determining the relationships of this rodent (see fig. 50). Stirton (1935) recognized the gross similarity of *Hystricops* to the castoroid beavers. Material available at that time did not permit any more definite conclusion than that the rodent was probably a castoroid. The unworn upper fourth premolar noted above was compared with an unworn upper fourth premolar of *Eucastor,* a small castoroid beaver also occurring in the Black Butte fauna. The unworn crowns of these two teeth are amazingly similar. In fact, one might suggest that the new material is merely a gigantic *Eucastor.* They also possess similarly developed stria and flexures, and the major cusps are identical in position and relative development. The stria of *Eucastor* are longer but are similar in their relative lengths to the new material. They are also similar in the development of roots

TABLE 7
MEASUREMENTS UPPER P4'S *Hystricops*
(IN MILLIMETERS)

	Greatest AP	T.	Bottom of hypos. to base of enamel	Length of stria			
				Hyp	Met	Para	Meso
F-5802	8.7	10.5	6.4	3.3			notch
F-20384	9.1	10.3	5.3	12.0	0.9	4.0	6.0

after eruption of the tooth. This is illustrated with both the Black Butte material and the Nevada material (Macdonald, 1956). Parallelism among rodents is rampant and can be demonstrated at many taxonomic levels. Some hystricomorph rodents closely parallel beavers in tooth evolution especially in development of full-length strias, hypsodont teeth, and other characteristics. Many examples of parallelism within lesser taxonomic groupings are evident. This is hard to recognize in worn teeth since the fundamental cusps are lost early in wear in the teeth of species with high-crowned teeth. The new material indicates that it is of a species that is definitely a member of the subfamily Castoroidinae. It further shows close affinities with *Eucastor* possibly with a common ancestor in *Monosaulax*.

As indicated above, it cannot be determined with the available material whether the Juntura and Nevada material is referable to *Hystricops* although the specimens compare closely on the basis of the characteristics seen.

Eucastor malheurensis n. sp.

Type: F-6683, a fragmental right lower jaw with P4-M2. Teeth in early stage of wear (fig. 43).

Referred material: F-5773, a lower left partial lower jaw with P4-M2 and a portion of the incisor, UO loc. 2341 (fig. 42).

TABLE 8
PREMOLAR MEASUREMENTS *Eucastor malheurensis*
(IN MILLIMETERS)

Specimen no.	Postero-anterior	Transverse	Height of crown
	P/4		
F-5848	4.8	3.5	7.0
F-5830	4.8	3.6	9.0
F-5840	4.8	3.6	11.8
F-5843	5.2	3.7	13.2
F-5554	6.0	4.4	11.9
F-5835	4.9	3.7	13.9
	P4/		
F-5556	4.5	4.6	15.2
F-5834	3.4	3.5	12.5
F-5832	3.7	4.0	12.0

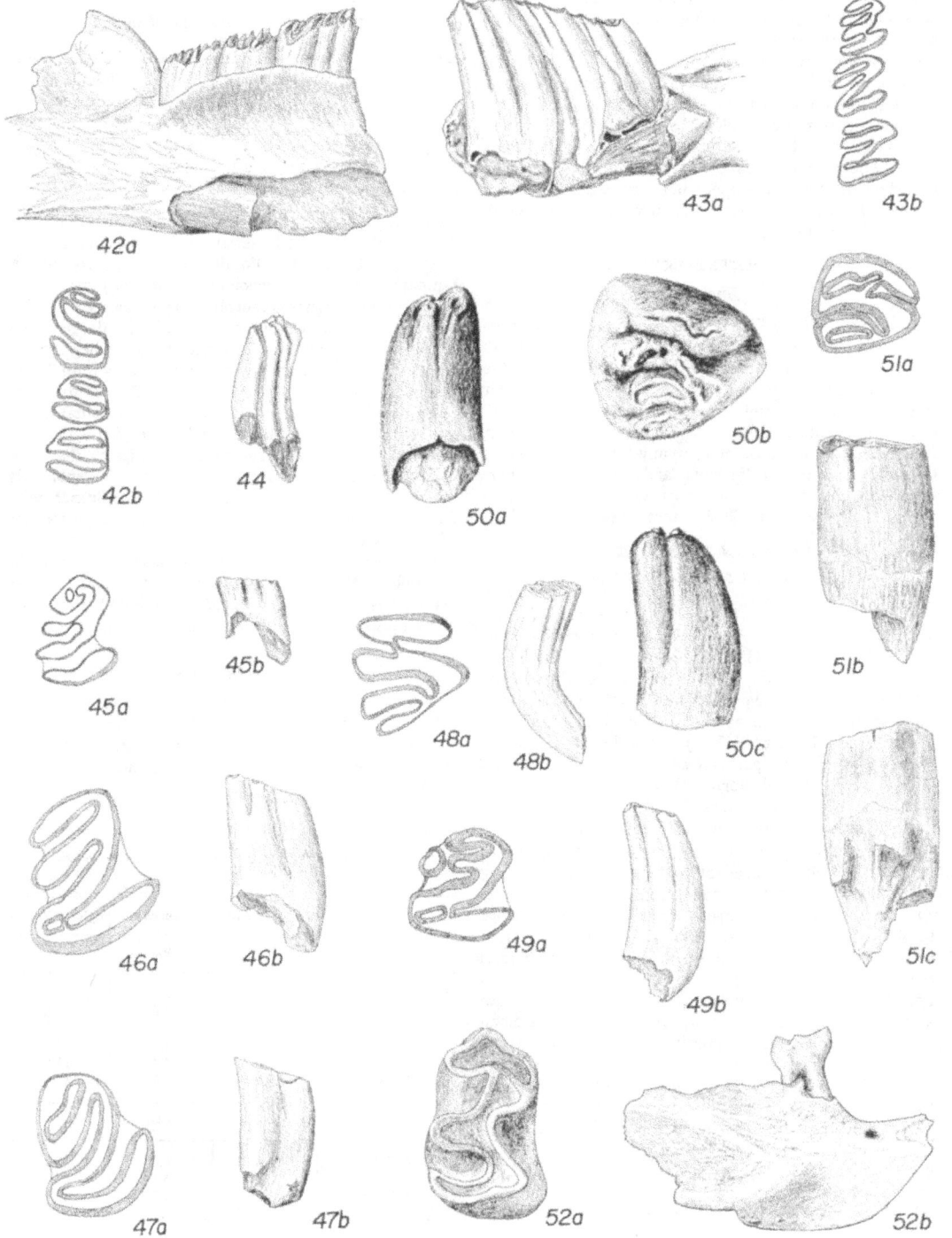

FIGS. 42–52. Rodentia of the Black Butte fauna.

Type locality: Juntura formation: Locality UO 2352. In the mouth of a small draw about one and one half miles west of Juntura, Oregon.

Diagnosis: About the size of *E. diversidens*. In the upper premolars the mesostria is longer than the parastria and the parastria longer than the metastria. The hypostria is essentially the full length of the tooth. In the lower premolars the mesostriid is longer than the parastriid. The teeth are prismatic.

Description of materials: *Lower dentition:* Fourth premolar largest member of the cheek-tooth series. The hypostriid is complete. The mesostriid is longer than the parastriid. In only one specimen was a metastriid in evidence and in this case (a little-worn specimen) it was only a notch. The metafossettid is persistent but is lost in the course of wear before the parastriid closes (fig. 46, 47). The paraflexid is generally straight although one specimen displays a hook at the distal end as seen in *E. lecontei*. The anterior and posterior faces of the tooth are nearly parallel. Flexids on all the teeth are markedly straight. In one specimen the metaconid is isolated from the tooth (fig. 49). This variation also occurs in *Dipoides stirtoni* (Shotwell, 1955: fig. 7C).

---

42. *Eucastor malheurensis* Shotwell and Russell, n. sp. Assigned specimen, left lower jaw, UOMNH F-5773, loc. 2341. Lingual view ×2.5 (42a). Occlusal view dentition P4-M2 ×2.5 (42b).
43. *Eucastor malheurensis* Shotwell and Russell, n. sp. Holotype UOMNH F-6683, loc. 2352. Right lower jaw, lingual view ×2.5 (43a). Occlusal view dentition P4-M2 ×2.5 (43b).
44. *Eucastor malheurensis* Shotwell and Russell, n. sp. Assigned specimen unworn upper P4, UOMNH F-10478. Labial view ×2.5.
45. *Eucastor malheurensis* Shotwell and Russell, n. sp. assigned specimen. Lower right DP4, UOMNH F-10479. Occlusal view ×5 (45a). Lingual view ×2.5 (45b).
46. *Eucastor malheurensis* Shotwell and Russell, n. sp. assigned specimen, lower left P4 little worn, UOMNH F-5554. Occlusal view ×5 (46a). Lingual view ×2.5 (46b). Mirror image illustration.
47. *Eucastor malheurensis* Shotwell and Russell, n. sp. assigned specimen, lower right P4 well worn, UOMNH F-5830, occlusal view ×5 (47a). Lingual view ×2.5 (47b).
48. *Eucastor malheurensis* Shotwell and Russell, n. sp. assigned specimen, upper right P4, UOMNH F-5834. Occlusal view ×5 (48a). Labial view ×2.5 (48b).
49. *Eucastor malheurensis* Shotwell and Russell, n. sp. assigned specimen, lower left P4, UOMNH F-5835. Occlusal view ×5 (49a). Lingual view ×2.5 (49b). Mirror image illustration.
50. *?Hystricops* sp. Unworn upper left P4, UOMNH F-20384, loc. 2332. Occlusal view ×3 (50b). Lateral views ×2 (50a & c).
51. *?Hystricops* sp. Well-worn upper left P4, UOMNH F-5802, loc. 2340. Occlusal view ×2 (51a). Lateral views ×2 (51b, 51c).
52. *Peromyscus* cf. *dentalis* Hall, lower right jaw fragment with M1, UOMNH F-18237. Occlusal view of M1 ×20 (52a). Labial view lower jaw ×10 (52b).

Upper dentition: The fourth premolar is the largest tooth of the cheek-tooth series. The mesostria is longer than the parastria which is in turn longer than the metastria. As in the lowers the hypostria is essentially the full length of the tooth. The hypoflexus and paraflexus tend to bypass each other (fig. 48). The metastria is a notch. The sides of the tooth do not taper. The flexi of all the upper teeth are rather straight. The flexi of all the upper teeth are rather straight.

Comparisons: The relative length of the strias and striids separates this species from other known species. The straightness of the flexi is similar to *E. lecontei;* however the new species has consistently longer stria and striids.

Family CRICETIDAE

Peromyscus cf. *dentalis* Hall (1930a)

A fragmental right lower jaw with M1 No. F-18237 is referred to *P. dentalis*. M1 is smaller than in *P. dentalis* but otherwise is assigned to that species. M1, Ap-1.2 mm., T-.7 mm. (fig. 52).

Order CARNIVORA

Family CANIDAE

Subfamily CANINAE

Vulpes sp.

Referred specimen: A talonid of a left M/1 F-5862 from locality UO 2337.

Description: Has a median groove and no posterior crest. The hypoconid is elongate and crested. The entoconid is small and separated from the metaconid by a smaller cusp. There is no cingulum present. The antero-posterior diameter of the talonid is 3.8 mm. and the transverse diameter 5.0 mm. The bicuspate talonid is a canid character. Its similarity in size and shape to that of *Vulpes* makes reference to this genus probable.

Subfamily BOROPHAGINAE

Osteoborus sp.

A partial associated skeleton (F-5938) was excavated in a small quarry at locality UO 2335. The specimen included: right humerus, right femur, fragment of left femur, pelvis, fragment of sacrum, right ulna, scapula, both scapholunars, metapodials, ribs, and vertebrae. There is a small entepiconular foramen present in the humerus, but no supratrochlear foramen; lesser tuberosity is short vertically, not extending down shaft; lateral distal trochlea strongly oblique and sharply crested along lateral border. The bones were evidently parts of one animal. There is close agreement with specimens of *Osteoborus diabloensis* Richey (1938). Positive generic determination cannot be made without information on the skull or dentition. The material is referred to *Osteoborus* (figs. 58, 59).

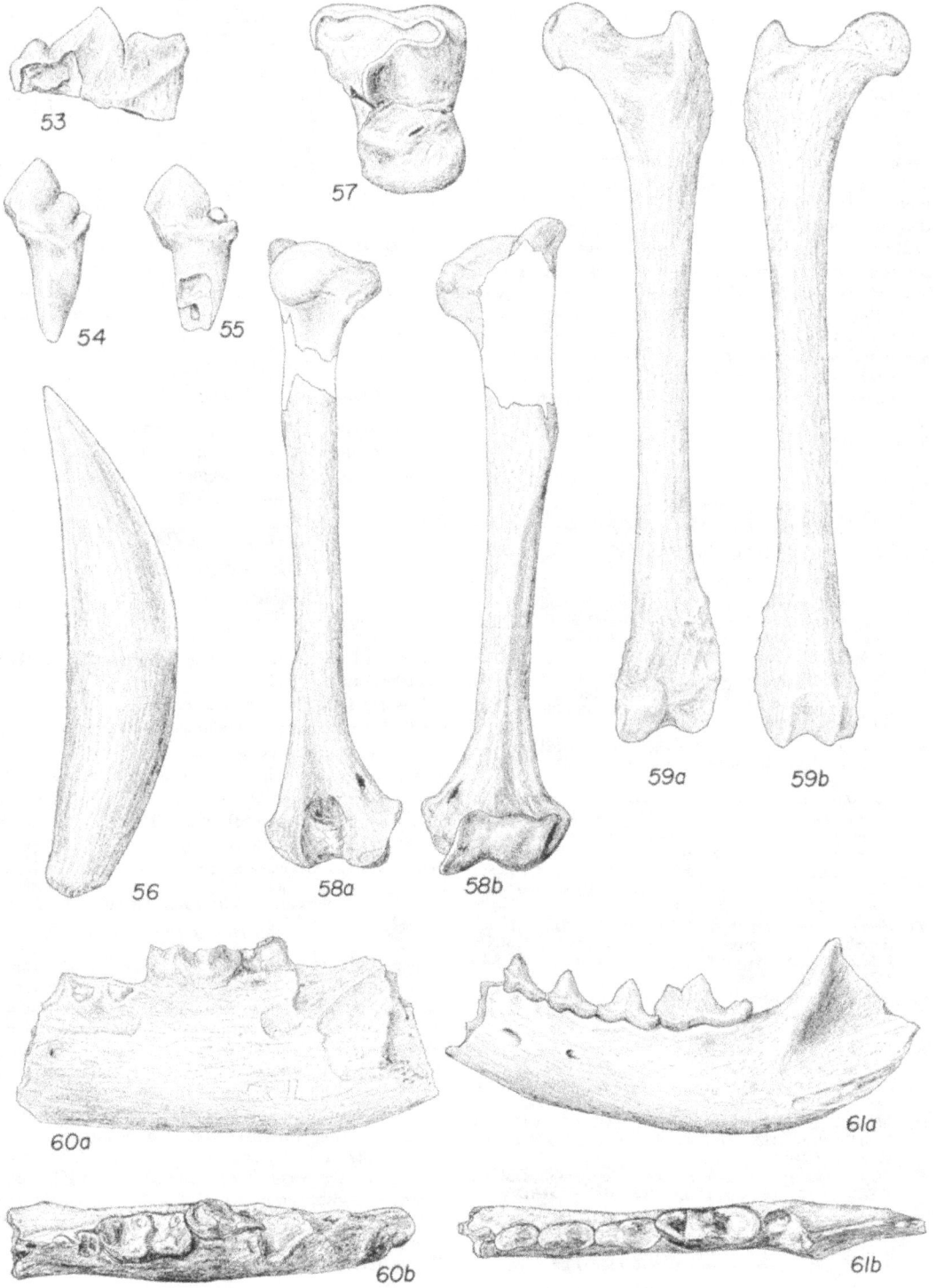

FIGS. 53–61. Carnivora of the Black Butte fauna.

TABLE 9

MEASUREMENTS OF *Osteoborus* SP. (IN MILLIMETERS)

Femur

Total length	183.0
Diameter through head and greater trochanter	40.5
Maximum diameter at midlength of shaft	14.0
Transverse diameter through distal trochlea	31.0

Humerus

Total length	165.0
Antero-posterior diameter of head	26.4
Transverse diameter of head at tuberosities	35.9
Maximum diameter at midlength of shaft	15.7
Greatest transverse diameter of distal end	37.0
Transverse diameter of distal trochlea	27.0

4th Metacarpal

Total length	50.0
Maximum transverse diameter through proximal end	9.0
Maximum transverse diameter through distal trochlea	8.4

Scapholunar

Transverse diameter	17.0
Antero-posterior diameter, including radio-palmer process	19.0
Vertical diameter not including radio-palmer process	10.0

Aelurodon sp.

Complete and fragmental teeth from localities UO 2343 and UO 2344 are assigned to *Aelurodon*. A calcaneum (F-6005) from locality UO 2348 is also assigned to *Aelurodon*. The material does not allow reference to species. See fig. 57.

FAMILY MUSTELIDAE

Martes sp.

An isolated lower first molar of a mink-sized mustelid is present in the collection, F-5942, locality UO 2334. The protoconid and metaconid are more widely separated than in the mink; however, the talonid basin is similar. The antero-posterior diameter of the tooth is 9.6 mm., width 4.0 mm. There is a distinct notch between the hypoconid and protoconid.

53. *Pseudaelurus* sp. Upper P4, UOMNH F-6575, loc 2343. Lingual view ×1.5.
54. *Pseudaelurus* sp. Upper P2 or P3, UOMNH F-6575, loc. 2343. ×1.5.
55. *Pseudaelurus* sp. Upper P2 or P3, UOMNH F-6575, loc. 2343. ×1.5.
56. *Pseudaelurus* sp. Upper canine, UOMNH F-6575, loc. 2343. ×1.5.
57. *Aelurodon* sp. Upper M1, UOMNH F-5607, loc. 2332. Occlusal view ×1.5.
58. *Osteoborus* sp. Humerus, UOMNH F-5938, loc. 2335. Posterior view ×0.75. (58a). Anterior view ×0.75 (58b).
59. *Osteoborus* sp. Femur, UOMNH F-5938, loc. 2335. Posterior view ×0.75 (59a). Anterior view ×0.75 (59b).
60. *Pliotaxidea* sp. Lower jaw with M1, M2, UOMNH F-11032, loc. 2344. Labial view ×1.5 (60a). Occlusal view ×1.5 (60b).
61. *Sthenictis junturensis* Shotwell and Russell, n. sp. Holotype, UOMNH F-6694, loc. 2344, left lower jaw. Labial view ×1.5 (61a). Occlusal view ×1.5 (61b).

Pliotaxidea sp.

A left lower-jaw fragment F-11032 from locality UO 2344 closely resembles *P. nevadensis* in characteristics of the M1. The specimen consists of a mandible broken at the ascending ramus and again just anterior to the P4 alveoli. Only the broken roots of the P4 are present. The first and second lower molars are present and complete. They are only moderately worn. There is a small foramina below the anterior root of the P4. The heel of the M1 lacks the external accessory cusps seen in *Taxidea*. It closely matches *Pliotaxidea nevadensis* in the orientation and placement of cusps on the first molar and in the occlusal outline of the tooth (see fig. 60). The specimen represents a badger much larger than indicated by the known *Pliotaxidea* material. In view of the size range of *Taxidea* it does not seem that size alone is a basis for separating the Black Butte specimen from *Pliotaxidea*.

Eomelivora sp.

A maxillary fragment with the alveoli for the fourth premolar and first molar (F-6250) indicates the presence of a large mustelid in the fauna. The alveolus of the fourth premolar indicates a narrow carnassial with a well-separated and prominent protocone. The molar was apparently narrow labially and broadened lingually. Comparison of the Black Butte material with an equivalent fragment of *Plesiogulo* including P4-M1 from McKay reservoir shows some differences in root arrangement. The McKay specimen has two well-separated roots at the labial side of the molar. The Black Butte specimen has a single small labial root. This difference suggests a narrower trigonid on this species similar to *Eomelivora*. This suggests a large mustelid with a prominent protocone on the upper P4 with a narrow M1. It is assigned to *Eomelivora* on this basis.

Sthenictis junturensis n. sp.

Type: Left lower jaw. Condylar portion missing. Canine, P1 and M2 missing. P2-M1 in place and complete. UOMNH F-6694, locality UO 2344, fig. 61.

Type locality: UO 2344. See map 3.

Diagnosis: A river otter about the same size as living forms but with much more slender teeth. Teeth simple lacking accessory cusps.

Description: P1 single-rooted; P2-P4 with two roots, not crowded, laterally compressed; only a trace of accessory cusp on P4; shear plane of M1 nearly antero-posterior, trigonid longer than talonid, metaconid strongly separate, talonid slightly basined, external margin separated from posterior edge of protoconid by distinct notch, internal margin without entoconid and continuous with posterior border of metaconid; M2 single rooted; two anterior mental foramina, one below middle of P2 and second below posterior root of P3; mandibular foramen placed ven-

TABLE 10

MEASUREMENTS OF TEETH OF *Sthenictis junturensis*
(IN MILLIMETERS)

	P2	P3	P4	M1
Length	5.7	6.6	7.7	12.7
Width	3.7	4.0	4.3	5.8

trally, near angle of jaw; masseteric fossa extends to point below middle of M2; depth of jaw, lingual side, below middle of M1 13.8 mm., below middle of P2 12.5 mm.

Comparisons: Gazin (1934) described two new species of *Lutravus?* from the Hagerman sites in Idaho. One is similar to the Black Butte specimen but is distinguished by a short jaw with the premolars crowded and reduced to three. In 1937, with additional material, Gazin referred the Idaho species to *Canimartes*. *C. cumminsii* Cope (1892) from the Blanco formation of Texas has a longer and narrower carnassial than that from Black Butte. Figures of the type (Cope, 1893) indicate a smaller and less distinctly basined talonid.

Lutravus halli Furlong (1932) from Thousand Creek, Nevada, is larger than the Black Butte specimen and the talonid is proportionally much shorter. Merriam, Stock, and Moody (1925) briefly described a fragmentary lower jaw from the Rattlesnake beds of Oregon which Furlong (1932) suggests is probably referable to *Lutravus halli*.

Mionictis incertus Matthew (1924) from Lower Snake Creek is near the Black Butte specimen in size but differs in carnassial characters. The talonid is longer with a high, crested hypoconid and a small cusp between the protoconid and hypoconid, the trigonid is shorter, and the premolars tend to be slightly more crowded. *M. elegans* Matthew (1924) is smaller and the M2 is double-rooted.

A mandible from the Ricardo formation, described by Merriam (1919) as *Mustela? buwalda*, is not as large as the one from Black Butte and has a large accessory cusp on the P4. The shear of the carnassial is more antero-posterior and the metaconid projects farther lingually.

Cernictis hesperus Hall (1935) from the Pinole formation, California, is differentiated by a relatively long P4. The talonid of the M1 is short and semi-basined with the hypoconid and entoconid poorly developed. A distinct ridge connects the metaconid with the tip of the protoconid.

Within the genus *Brachypsalis* only *B. matutinus* Matthew (1924) from Lower Snake Creek, and *B. angustidens* Hall (1930b) from Pliocene Kern River beds resemble the Black Butte species. *B. angustidens* is smaller and is distinguished by a shorter, deeply basined talonid. The premolars of *B. matutinus* are reduced and crowded; the hypoconid of the M1 is poorly developed.

Sthenictis dolichops Matthew (1924) from the Lower Snake Creek beds is much larger than *S. junturensis*. The cheek teeth are well spaced and relatively blocky. M1 is similar but more massive; the hypocone is not as near the external border. The Black Butte species closely resembles *S. bellus* Matthew (1932), from Sheep Creek "horizon A," in shape and spacing of the cheek teeth and is only slightly smaller. P4 of *S. bellus* has a strong accessory cusp and could represent an ancestral form to both *S. dolichops* and *S. junturensis*. *Sthenictis robustus* (Cope, 1890) was based on a lower jaw from the Blanco formation of Texas. This species appears to be related to *Brachypsalis* and *Canimartes*, but more material is needed before these three genera can be clearly distinguished. Available evidence indicates that the Black Butte specimen is most closely allied to species of *Sthenictis*.

Mustela palaeosinensis Zdansky (1924) from the Pliocene of China, is similar to *Mustela? buwalda* from the Ricardo and resembles species of *Sthenictis*.

The lower carnassial of *Sinictis dolichognathus* Zdansky (1924), also from the Pliocene of China, has a reduced metaconid and a relatively large bulbous hypoconid. The mandible is markedly shallower than in most mustelids.

Pannonictis pliocaenica Kormos (1931) from the late Pliocene of Hungary is larger than the Black Butte specimen and bears many affinities with the grisonines. The paraconid of M1 is low and the talonid deeply basined.

An indeterminate mustelid mandible reported by Zdansky (1927) from the late Miocene of China is referred to *Pannonictis* by Kormos (1931). Zdansky's specimen, although lacking the carnassial, is similar to the Black Butte jaw.

FAMILY **FELIDAE**

Pseudaelurus sp.

A number of complete and fragmentary teeth and skeletal elements of the cat *Pseudaelurus* were collected at locality UO 2343. These fragments apparently represent a single individual. Present are upper right canine, upper left P4, lower left and right P4's and less complete fragments of other dental elements. Also present were right ectocuneiform, right navicular, fragmental calcanium and metapodials, tibia and vertebrae fragments, distal end of the tibia and pelvis fragments, all specimens F-6675.

The canine is complete and is 62 mm, long from root tip to crown tip. At the base of the enamel the tooth measures 14 mm. antero-posteriorly and 9.3 mm. transversely. It is laterally compressed and much flatter on the internal side than the external. The anterior and posterior faces have a sharp edge. The posterior edge is drawn into a small sharp ridge. This ridge has minute transverse ridges but is not serrate.

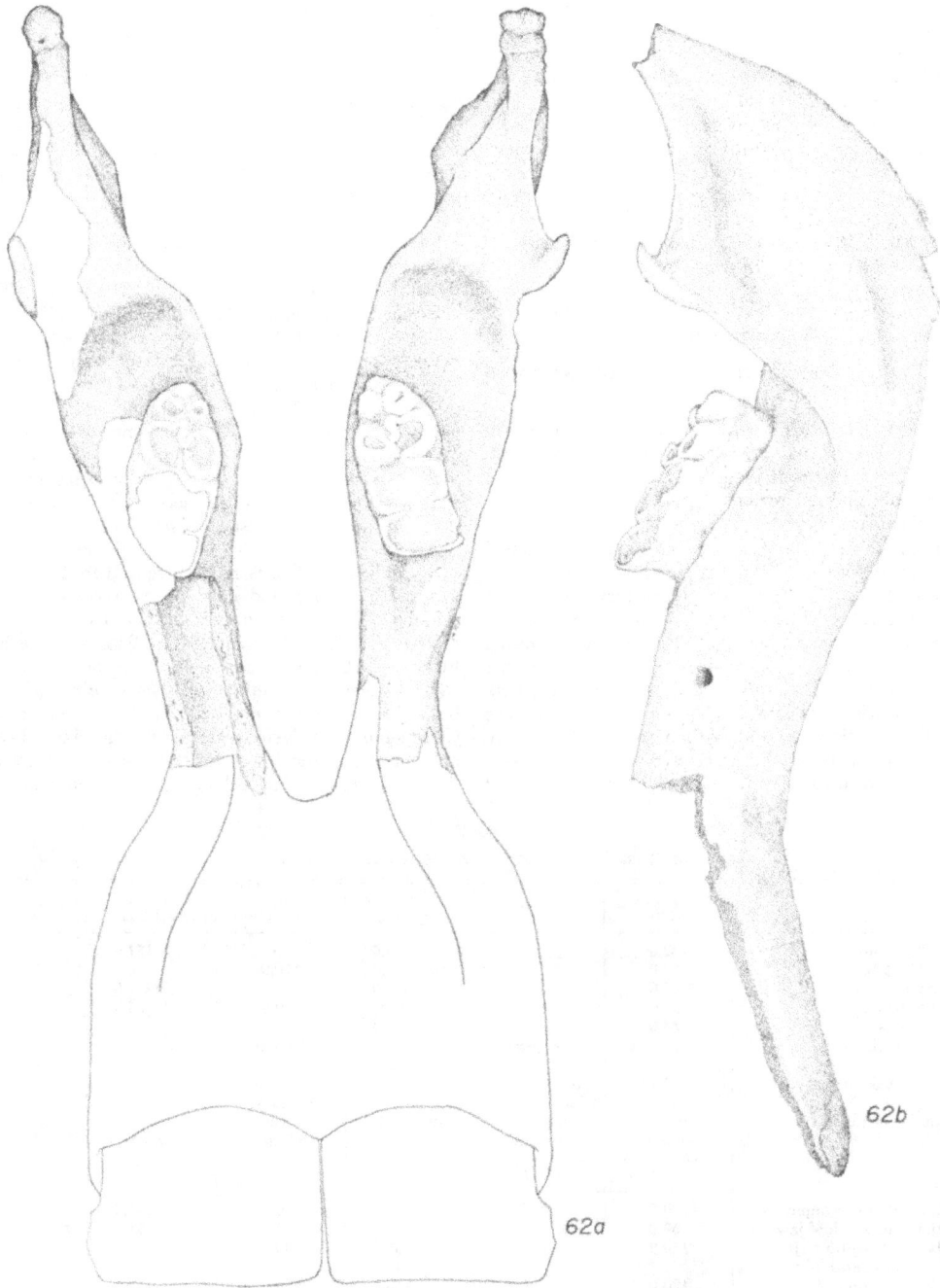

FIG. 62. *Platybelodon* (*Torynobelodon*) cf. *barnumbrowni* Barbour, mandible UOMNH F-7978, loc. 2340. Dorsal view ×0.25 (62a). Lateral view ×0.25 (62b).

The upper fourth premolar lacks the protocone but is otherwise complete. There is a strong pre-parastyle. The parastyle itself is rather small. The carnassial notch is narrow. The lower fourth premolars lack the anterior portions of the teeth. The accessory cusp is strong with a very strong cingular cusp immediately posterior to it. The small parastyle on the upper fourth premolar, the broad heel of the lower fourth premolar, the narrow carnassial notch of the upper fourth premolar suggest affinities with *Pseudaelurus* rather than *Nimravus*. The available material does not allow a species assignment (figures 53–56).

Order PROBOSCIDEA

Family GOMPHOTHERIIDAE

Platybelodon (Torynobelodon) cf. *P. barnumbrowni*
Barbour, 1931

Locality UO 2326 produced a right mandible, F-7975, with M2 and M3; left lower tusk fragment, F-7977 and a fragmental upper tusk and locality UO 2340 a right and left mandible, F-7978, each with M3 (figs. 63–67).

The lower tusk is concave dorsally, with the labial edge higher posteriorly. The tusk is encircled by corrugated enamel. The grooves are more prominent on the ventral side where three main ribs rise above the generally convex surface. No internal dentinal tubules are present. Length of the tusk is (basal part missing) 465.0 mm. There is no indication on the tusk of the position of alveolus border. The sides are parallel to within 110.0 mm. of the tip. It is 132.0 mm. wide. The flange at the anterolateral edge increases the width to 155.0 mm. The height of the lingual edge is 52.0 mm. with a slight concavity along the vertical axis. The lateral edge is rounded (see figs. 64, 65).

The upper tusk is ellipsoidal. It is thick dorsally, tapering ventrally, with moderate down-curving. It is encircled with corrugated enamel. The length of the fragment is 255.0 mm.; the extreme tip is missing. There is no indication of the position of alveolar border (figs. 66, 67).

The M3 is simple with single trefoiling. The ectotrefoils have simple anterior and posterior spurs. The slight anterior spurs are present or absent on internal hemilophid or metalophid and tritolophid. There are broad, transverse valleys blocked at the mid-base by the expanded trefoils. No central conules are present. The molars have traces of cement and are low crowned without cingula.

The mandible, F-7978, is short, stocky, and lacking condyles. The symphysis is incomplete. The masseteric fossa is small and slightly impressed. The ascending ramus is low and slightly crushed in fossilization. The alveoli of the second molar are nearly filled in with bony tissue. The third molar is well worn.

Mandible, F-7975, is much larger than F-7978; the symphyseal region and most of the second molar are missing. The third molar is little worn.

The mandibles, F-7975 and F-7978, and the lower tusk, F-7976, closely resemble the type specimen of *Platybelodon (Torynobelodon) barnumbrowni* (Barbour, 1931: see figs. in Barbour, 1932) from the late Clarendonian of Nebraska. The large jaw, F-7975, and the upper and lower tusks were found associated. They are regarded as belonging to one individual.

TABLE 11

JAW AND TUSK MEASUREMENTS OF *Platybelodon* (IN MILLIMETERS)

	Lower tusk F-7976	Upper tusk F-7977	*P. grangeri*	*P. donovi*	*P. (T.) barnumbrowni*	*P. (T.) loomisi*
Width at insertion	132.0		168.0		155–157.0	
Width at tip	155.0		163–166.0	110.0	158.0	114.0
Length	°500.0		360–510.0		°559.0	
Thickness, average	33.0			30.0	25.0	
Thickness, maximum	33.0		32.0			
Upper-tusk enamel	completely encircling	completely encircling	lateral band	lateral band		
Lower-tusk enamel	completely encircling	completely encircling		absent on up. surface		
Interior of tusks	dentinal laminae	dentinal laminae	dentinal tubules	dentinal tubules	dentinal laminae	dentinal tubules
	Jaw, F-7978	Jaw, F-7975				
Min. width of symphysis	180.0	220.0	165.0	130.0	°191.0	
Minimum width of jaw	69.0	82.0			104.0	
Maximum width of jaw	120.0	145.0	106.0	90.0	139.0	
Maximum depth of jaw below middle of M/2	120.0	145.0	178.0	120.0	155.0	
Number of crests, M/3	/4½	/4½	5–6/5½–6¼	4½/5 1/3	/4½	
°=estimation.						

TABLE 12

TOOTH MEASUREMENTS OF *Platybelodon* (IN MILLIMETERS)

	Jaw, left	F-7978 right	Jaw, F-7975	P. grangeri	P. donovi	P. (T.) barnumbrowni
M/3 length	147.0	155.0	179.0	192–205.0	168–170.0	165–202.0
width (3rd lophid)	69.0	72.0	77.0	68– 69.0	67.0	80.0
height, int. (3rd lophid)	40.0	32.0	49.0	69– 70.0		29.0
height, ext. (3rd lophid)	31.0	27.0	48.0			54.0
index (W/L×100)	47.0	46.5	43.0	57.0	40.0	44.0
M/2 length			ᵃ107.0	135.0		105.0
width, maximum			ᵃ70.0	63.0		69.0
height, max. int.			ᵃ25.0			19.0
height, max. ext.			ᵃ35.0			54.0
index (W/L×100)			ᵃ65.0	46.5		65.5

Absence of a flange on the lower tusk of *P. barnumbrowni* is probably due to greater wear in an older animal. Wear of M3 in this specimen has obliterated the trefoil pattern from the protolophid and metalophid which hinders comparisons with the less-worn Black Butte specimens. Following Vanderhoof (1937), enamel was determined with a microscope and polarized light. No mention was made of tusk enamel in the description of *P. barnumbrowni*.

Platybelodon donovi Borissiak (1927), from the middle Miocene of the North Caucasus, has a much longer and more slender symphysis than the Black Butte form but the molar trefoil pattern is similar.

The late Miocene *P. grangeri* Osborn (1929) from Mongolia has a long, slender jaw that flares into a short, deeply basined scoop. The molars are high crowned. Dentinal tubules are present in the lower tusks of *P. donovi* and *P. grangeri*.

Amebelodon fricki Barbour (1927), a late Pliocene mastodont from Nebraska, is distinguished by its extremely elongated, slender lower tusks.

Only lower tusks and the anterior part of the symphysis of *Serbelodon burnhami* Osborn (1933), from the early Pliocene of California, have been found. The wide posterior symphyseal region is suggestive of *Torynobelodon* as is the shape of the lower tusk and the absence of dentinal tubules. Surface grooves appear to be present in the figure but are not mentioned. The tusks are slightly smaller than those from Black Butte.

P. (T.) loomisi (Barbour, 1929) from the middle Pliocene of Nebraska, is known from a single lower tusk. It is narrower than *P. (T.) barnumbrowni* but agrees in curvature and grooving. Dentinal rods are present but Gregory (1945: 481) and others doubt their taxonomic significance.

The large jaw, F-7975, is of a young adult, and the mandible, F-7978, is of a fully mature, but smaller, adult. This difference could be sexual.

FAMILY **MAMMUTIDAE**

Mammut (Pliomastodon) furlongi n. sp.

Type: UOMNH F-10291, a mandible, nearly complete with M1-M3 present on each side (fig. 68).

Referred specimens: UOMNH F-6208, a nearly complete upper third molar, locality UO 2343 (fig. 69).

Type locality: Juntura formation quarry 3, UO 2448.

Diagnosis: A mastodon similar to *P. matthewi* in stage of evolution and size. Narrow deep symphyseal trough. Lower and upper M3 with numerous accessory cusps. Trefoils are present but poorly developed on labial side in lower M3, lingual side on upper M3. Heavy anterior cingulum on upper and lower M3's. Strong posterior cingulum on lower M2. M1 and M2 lack accessory cusps. Lower M3 has four crests and a poorly developed partial fifth. Upper M3 has three crests and a partial fourth. There are no lower tusks. No lingual or labial cingula. Named in honor of Eustace L. Furlong who made early collections in the western portion of the Juntura basin.

Description of material: *Lower dentition:* M3 four complete lophs and a partial fifth are present. Partial loph divided by antero-posteriorly directed valleys into three poorly separated parts. Trefoils are present but not prominent until late wear on labial side of tooth. Accessory cusps numerous on crests of lophs, about six to each loph. Valleys are relatively clean. Anterior cingulum strong and ridges with cuspids. Cingulum rises medially. No lingual or labial cingula.

M2, three equal lophs. Trefoils present but only poorly developed, and only on labial side. No accessory cusps are present on trefoiled cusps. Posterior cingulum heavy and dotted with accessory cusps. Anterior cingulum small, no labial or lingual cingula. Valleys are clean.

M1, trilophodont and much smaller than M2. Trefoils shows poor development. No anterior cingula, posterior cingulum weak with no labial or lingual cingula present. No accessory cusps in evidence.

Mandible rather short and heavy; this may reflect the young adult age of the individual. There is a deep

63a

63b

64a

64b

67

65

66

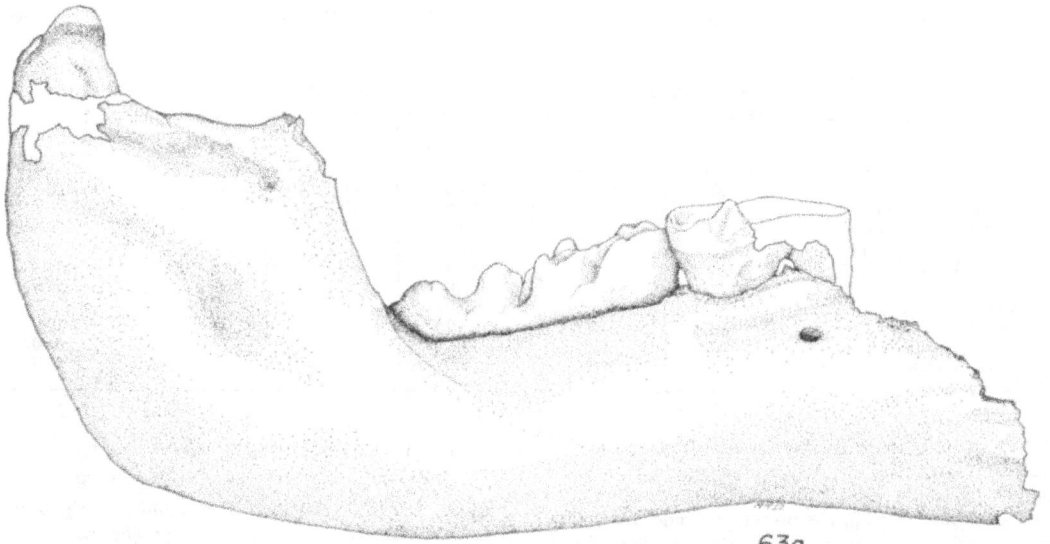

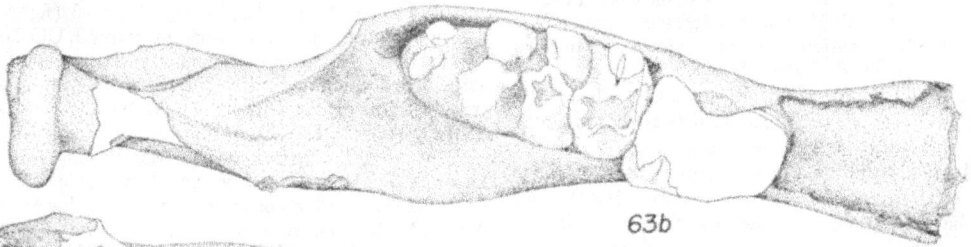

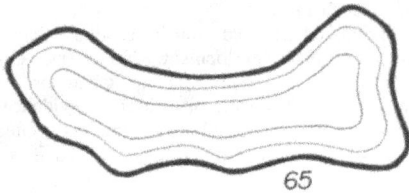

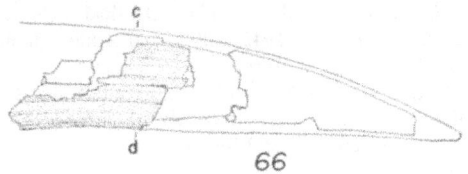

FIGS. 63–67. *Platybelodon.*

narrow symphyseal trough. Probably extended considerably farther out as indicated in illustration (fig. 68). There is a large foramen below the anterior portion of the second molar. A smaller one is seen near the anterior end of the portion of the trough present in the type specimen. There were no tusks present or evidence of alveoli for any.

Upper dentition: The only material of the upper dentition assigned to this species is a nearly complete upper third molar collected a few yards east of quarry 3, the type locality. It is from similar sediments and probably the same lentil. It is definitely not referable to the other proboscidean from this fauna.

The third upper molar narrows abruptly posteriorly. It has three lophs and one partial loph. Trefoils are rather prominent on the lingual side. The partial loph consists of a cluster of five unequal cusps. Valleys are clean. The anterior cingulum is strong with few accessory cusps. It extends around the edge of the tooth to the lingual side. However, there is no lingual or labial cingula other than this extension.

Comparisons: *P. matthewi* of the Great Plains does not have the development of accessory cusps as seen in the Black Butte species. The upper and lower third molars of *P. matthewi* are larger; development of the last partial loph is similar suggesting similar stages of development of the two species. *P. americanus praetypica* of Europe has a much heavier partial loph than *P. furlongi* in lower third molars as well as a more poorly developed anterior cingulum. *P. sellardsi* of Florida is about the same size as the material from Black Butte. However, it possesses lower tusks and has generally wider teeth. A partial fifth crest on the lower third molars is similar in development to the new species. *P. vexillarius* of California has four full crests on the upper third molar which is also broader posteriorly. The partial last crest of the lower third molar is much more prominent than in the Black Butte species. *P. nevadensis* from Nevada differs from the new species in the appearance of four complete crests on the upper third molar and also the presence of a prominent lingual cingula. They are, however, similar in size. *P. cosoensis* of California also has four complete lophs in the upper third molar and few accessory cusps. There is a strong cingulum present on the lower third molars. The symphyseal trough is wide and

---

63. *Platybelodon (Torynobelodon)* cf. *barnumbrowni* Barbour, right lower jaw, UOMNH F-7975, loc. 2326. Lateral view ×0.25 (63a). Dorsal view ×0.25 (63b).
64. *Platybelodon (Torynobelodon)* cf. *barnumbrowni* Barbour, lower tusk, UOMNH F-7977, loc. 2326. Ventral view ×0.25 (64a). Lateral view ×0.25 (64b).
65. *Platybelodon (Torynobelodon)* cf. *barnumbrowni* Barbour. Cross-section view at *a* on figure 64b.
66. *Platybelodon (Torynobelodon)* cf. *barnumbrowni* Barbour, fragmental upper tusk, UOMNH F-7977, loc. 2326. Lateral view ×0.25.
67. Cross-sectional view of upper tusk of figure 66 at *c-d*.

TABLE 13

MEASUREMENTS OF *Mammut (Pliomastodon) furlongi*
(IN MILLIMETERS)

		Left	Right
Type mandible—UOMNH F-10291			
M3	AP	167	167
	T ant. loph	74	71
	T post. loph	68	—
M2	AP	105	107
	T ant. loph	57	56
	T post. loph	67	67
M1	AP	73	73
	T ant. loph	42	45
	T post. loph	52	55
Depth of jaw below M1		124	128
M1-M3		342	333
M2-M3		274	270
M1-M2		179	177
Depth of valleys on lingual side of M3, Top of cusp to base of valley			
Between first and second lophs		29	30
Between second and third lophs		29	30
Between third and fourth lophs		24	—

Referred upper third molar UOMNH F-6208

AP	148
T at first loph	89 (estimated)
T at second loph	82
T at third loph	66
T at last partial	38

Crest to valley between second and third loph—27

rather short. The partial loph of the lower third molar is well developed towards *M. americanus*. The Black Butte species differs in all these respects.

The new species differs from each of the known species in several or all of the following characteristics: (1) Development of accessory cusps on upper and lower third molars. (2) Long narrow symphyseal trough. (3) Poorly developed fifth loph on lower third molars. (4) Incomplete fourth loph on upper third molars. (5) Heavy anterior cingulum on lower third molar. (6) No lingual or labial cingula. (7) No lower tusks present.

Mammutidae

A tusk fragment with an enamel band 41 mm. wide (F-11293) indicates a possible third proboscidian in the fauna. Although the fragment is only a small section not much wider than the enamel band and about 50 mm. long, the curvature indicates a large diameter for the tusk. A third proboscidian is indicated since it is not expected that a two-tusked mastodon will have enamel bands on the tusks.

A number of foot and limb elements of proboscidians are in the collection. They are not assigned to any of the known species from the fauna except where they are in association with dental elements.

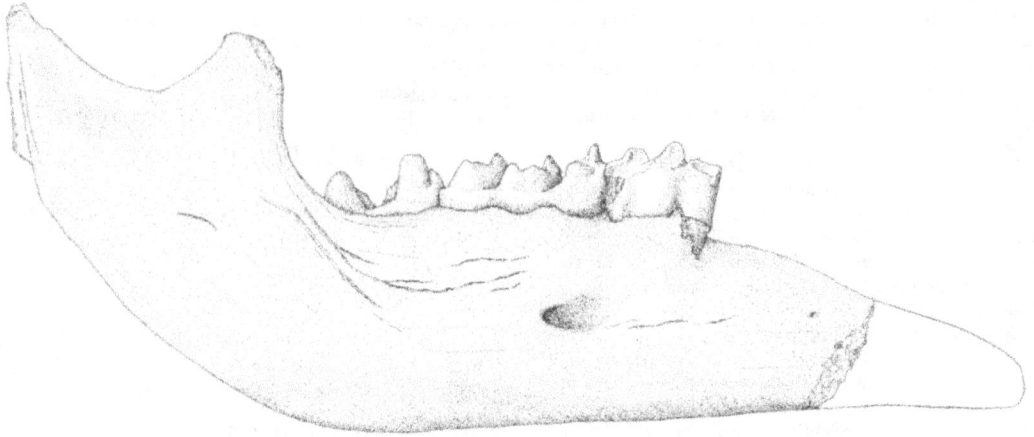

68a

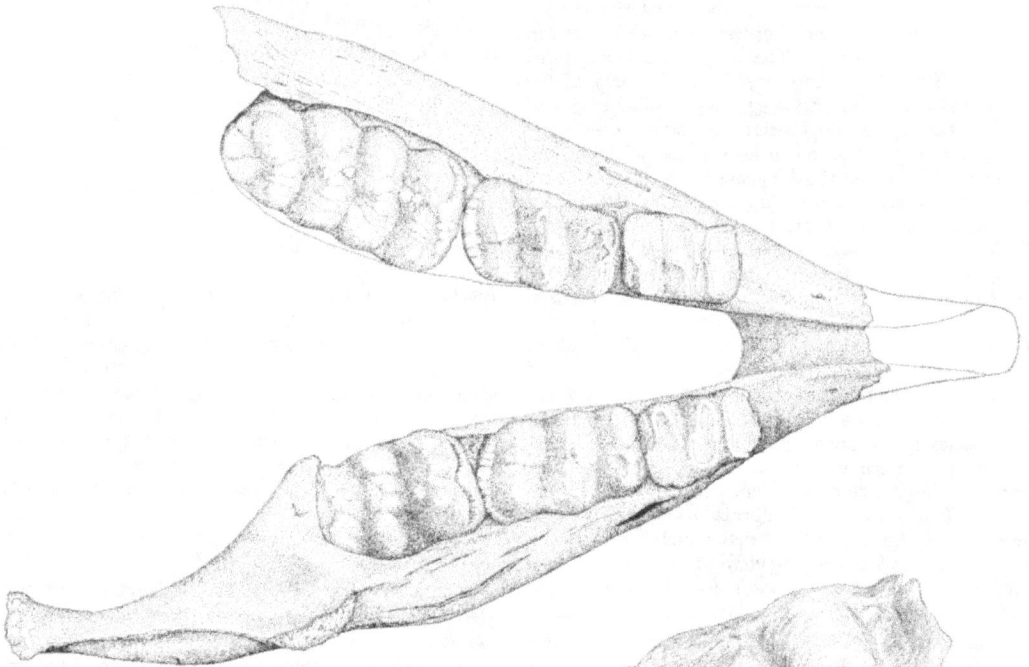

68b

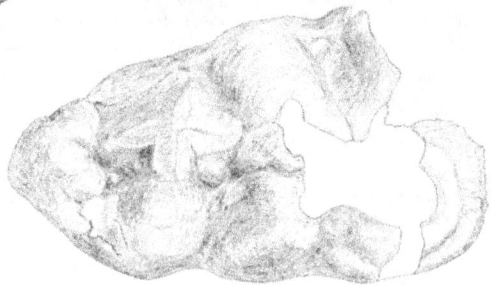

69

FIGS. 68–69. *Mammut.*

ORDER **PERISSODACTYLA**

FAMILY **EQUIDAE**

Hipparion cf. *condoni* Merriam (1915)

Horse material is common in the collection. Nearly every locality has some horse material from it although never as abundant as camel. Quarry 3 produced several nearly complete lower jaws and numerous other loose teeth and skeletal elements. Material from this and other localities in the Juntura formation includes sixty-five isolated teeth, fragments of scapula, pelvis, a number of astragali, fragments of tibia, calcanii, radii, humeri, metapodials, a complete scapula, and nearly complete tibia. An associated partial upper dentition is also included (fig. 70, 71).

A cursory examination of the available material leads the worker to suggest that *Hipparion, Nannippus,* and *Neohipparion* are all present. However, careful study of the associated material, which represents different stages of wear, and the comparison of homologous teeth, among the isolated specimens, leads to the conclusion that this material represents a single species of horse. The skeletal elements indicate that a horse of only one size is present. Variations within the dentition and changes in the course of wear, seen as isolated examples, could be confused with other genera which approach the characteristics displayed. Since this material represents the only well-known early Pliocene, Great Basin hipparion horse, as far as the extent of variations go, it seems likely that examination of horse materials from other Clarendonian faunas in the Great Basin might show errors in the original designation of generic affinity. Because of this possibility the variations encountered and the associated materials available are illustrated. A somewhat different method of describing tooth differences and variations due to wear is used here. The variations seen in the characteristics, usually of significance, are described separately rather than the individual teeth as whole units. This is not new but seems more apt in this instance since it is the difference between the expression of characteristics from tooth to tooth in the series and in the course of wear that can be misconstrued as species or even generic differences in less complete material.

Upper dentition: The available material consists of an associated partial upper dentition and thirty isolated premolars and molars. The protocone tends to be flattened on the lingual side. It is often lenticular in little-worn teeth becoming rounder with wear (figs. 72, 73, 74). In the most worn teeth there is no sign of

attachment of the protocone to the protoconule, even on the second premolar where this usually occurs first. Para-, meso-, and metastyles are present. The metastyle is smallest but sharp. The parastyle is sometimes grooved. The mesostyle is narrow but high, usually the highest (figs. 78, 79). The hypostyle is often present in little to moderately worn specimens. The protostyle is often present giving a sharp exterior corner to the protoconule. The pli caballin has from one to three folds, with two folds on all the second premolars seen. The third and fourth premolars usually have a single fold, while the first and second molars have two folds or one highly complicated fold. The third molar has as many as three folds (figs. 70–73, 78, 79). The hypocone is lenticular to rounded. The postprotoconal valley is sometimes much deeper than the proprotoconal valley on the M1 and 2. It is so deep in some specimens that it joins the prefossette (see fig. 73). This connection is probably broken early in wear and is no doubt the origin of extra or isolated loops which are sometimes seen on or adjacent to the prefossette posterior border.

In early wear the posterior border of the postfossette and the anterior border of the prefossette are incomplete and open. The posterior border of postfossette closes after the anterior border of the prefossette in the course of wear. The postfossette and prefossette are connected in early wear on second premolars. Fossette borders are complicated; but the outer borders are not as complicated as the inner. There is at least one deep plication (pli hypostyle) on the posterior border of the postfossette in well-worn specimens with as many as two or three plications on less-worn material. A varying number of smaller plications may also be present on little-worn material. The anterior border of the prefossette always has one large pli protoloph with as many as five additional plications of varying depth in little-worn specimens. Little-worn specimens show the teeth to be longer (antero-posteriorly) than wide (transversely) measured at the occlusal surface and nearly square to wider than long in well-worn specimens or at the base of little-worn material. Measurements from a little-worn fourth premolar, F-5870, at the occlusal surface are AP-23.2 mm., T-18.5 mm. Measurements at the base of the hypoconal groove are AP-17.3 mm., T-19.5 mm. The height of crown of this specimen is 41.5 mm. It is as high as any material seen, and still shows wear. The changes in the occlusal outline of the tooth with wear, demonstrated above, can, along with variations in occlusal pattern, lead to misidentifications (see figs. 70, 71). For instance, little-worn teeth are very similar to species of *Neohipparion,* moderately worn specimens are more likely to be assigned to *Hipparion,* and well-worn material resembles *Nannippus,* especially *N. tehonensis.* The availability of associated dentitions or partial dentitions is the only way errors can be avoided.

---

68. *Mammut (Pliomastodon) furlongi* Shotwell and Russell, n. sp. holotype, UOMNH F-10291, loc. 2448, lower jaw. Lateral view right side ×0.25 (68a). Dorsal view ×0.25 (68b).

69. *Mammut (Pliomastodon) furlongi* Shotwell and Russell, n. sp. referred specimen, UOMNH F-6208, loc. 2343. Upper M3 occlusal view ×0.25.

TABLE 14

MEASUREMENTS OF *Hipparion condoni*

Left lower F-10283—little worn   Right lower F-11020—well worn

Measurements of teeth at occlusal surface

P2-M3	145 mm.	140 mm.
P2, Ap-	26.1 mm.	
T-	10.8 mm.	
P3, Ap-	23.7 mm.	
T-	11.5 mm.	
P4, Ap-	22.8 mm.	22.3 mm.
T-	10.5 mm.	12.6 mm.
M1, Ap-	20.8 mm.	20.1 mm
T-	9.1 mm.	9.5 mm.
M2, Ap-	22.2 mm.	20.3 mm.
T-	8.8 mm.	10.3 mm.
M3, AP-	little worn	24.2 mm.
		9.3 mm.

Depth jaw below M2—60 mm.
(labial)   P4—52 mm.
           P2—43 mm.

Posterior border of symphysis to posterior edge—265 mm.

Metacarpal—F-5900

    Total length—215 mm.
    Shaft (center)—width—28 mm.
              thickness—22 mm.
    Width articular surface distal end—33.5 mm.
    Width proximal end—38.2 mm.

Tibia

    Lacking only surface of proximal end to be complete—F-10796
    Length—30.8 mm.
    Least width of shaft—37.3 mm.
    Thickness of shaft at that point—25 mm.
    Width distal end—57.7 mm.

Astragali	Total length	Between centers of ginglymi
F-5589	48.8 mm.	22 mm.
F-5523	51.6 mm.	23.2 mm.
F-6023	56.8 mm.	21.3 mm.
F-5526	53.3 mm.	25.2 mm.

Hoof

    F-11241
       Width articular surface 34.5 mm.   Total length—45.1 mm.
       Depth of anterior notch 14.4 mm.

Lower dentition: The available material includes a lower left jaw, nearly complete, with P2-M3, F-10283. The dentition of this specimen is little worn, with the third molar just erupting to wear. A lower mandible with left and right jaws complete, F-11020, lacks incisors; P4-M3 are present on the right side, and first molar on the left side. This dentition is moderately worn. A deciduous dentition in a jaw fragment, F-10285, is unworn. Thirty-three isolated premolars and molars are also present (figs. 75, 77). In the lower dentition the parastylid is strong and sharp on all teeth except the second premolar. The pli caballinid

is present with one fold on little-worn premolars. It appears as a ridge on the hypoconid side of the groove in more worn premolars and on molars of all stages of wear. The metaconid and metastylid vary greatly from tooth to tooth in the cheek series (see fig. 75), and with wear. On the second premolar they are unequal in size, the metastylid being much larger. This relationship is exaggerated with wear. On the third and fourth premolars they are round to ovoid and equal in size. On the molars they are round and equal. The posterior lingual border of the metastylid is angular to squared off in worn specimens. The metaconid-metastylid groove is narrow and shallow on the second premolar. On the third premolar it is wide and shallow with the cusps well separated. On the fourth premolar and first two molars it is open, *U*-shaped, especially in worn specimens. The groove narrows from the top to the bottom of the tooth column. On the third molar they are well separated by a broad shallow groove. The entoconid is large or larger than the metaconid or metastylid. It has a sharp anterolabial corner, almost a stylid on little-worn teeth but becomes rounder with wear. The hypoconid forms a prominent stylid (hypostylid) on the lingual side of the tooth on all teeth. A plication is often present in the metaflexed or at the base of the metaconid on little-worn teeth. This is not so common in worn teeth. Only one example of a plication at the base of the metastylid was found. Plications on the labial border of the entoflexid are rarely present. The paralophid is a well-developed and prominent feature on the lingual side of the tooth. It is stylidlike in character. The well-developed hypostylid and paralophid, combined with the entoconid, metastylid, and metaconid, give the lingual side of the dentition a "picket fence" appearance.

Comparisons: The type specimen of *Hipparion condoni*, consists of a well-worn P4-M1. It was compared with equally worn material from the Black Butte localities. There are no differences between occlusal patterns of these materials. However, *H. condoni* has stronger roots than the Black Butte species. The fact that the type specimen of *H. condoni* is well worn could hide differences present in less-worn teeth.

70. *Hipparion condoni* Merriam, partial upper right dentition UOMNH F-5870, loc. 2338. Lateral view ×2 (70*a*). Occlusal view ×1.5 (70*b*).
71. *Hipparion condoni* Merriam, upper M1 UOMNH F-11069, loc. 2340. Labial view ×2 (71*a*). Occlusal view ×1.5 (71*b*).
72. *Hipparion condoni* Merriam, little worn upper ?M1 UOMNH F-10935, loc. 2344. Occlusal view ×1.5.
73. *Hipparion condoni* Merriam, moderately worn upper M1, UOMNH F-5618, loc. 2332. Occlusal view ×1.2.
74. *Hipparion condoni* Merriam, well worn upper M1, UOMNH F-5636, loc. 2332. Occlusal view ×1.5.

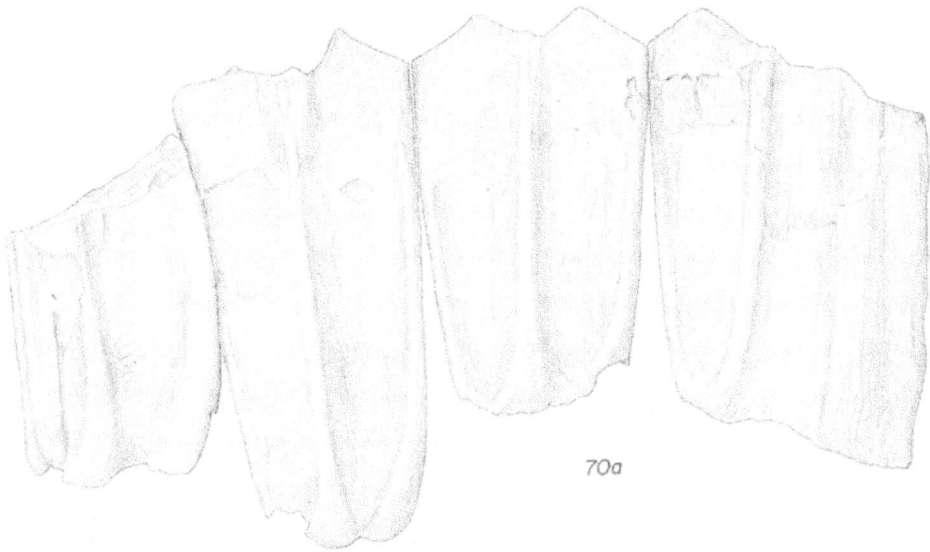

70a

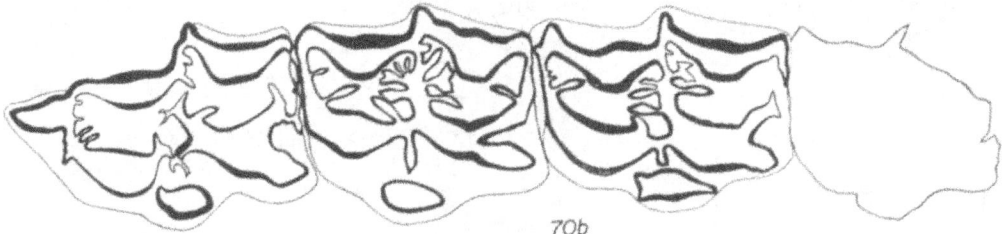

70b

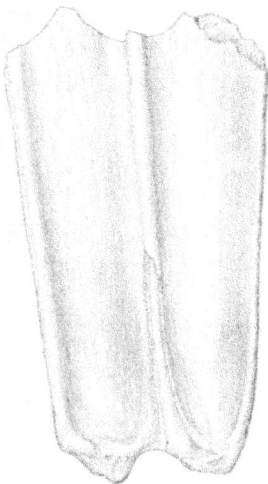

71a

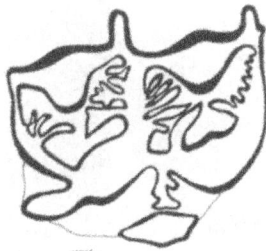

71b

72

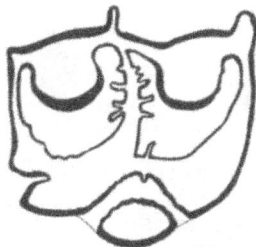

74

73

Figs. 70–74. *Hipparion.*

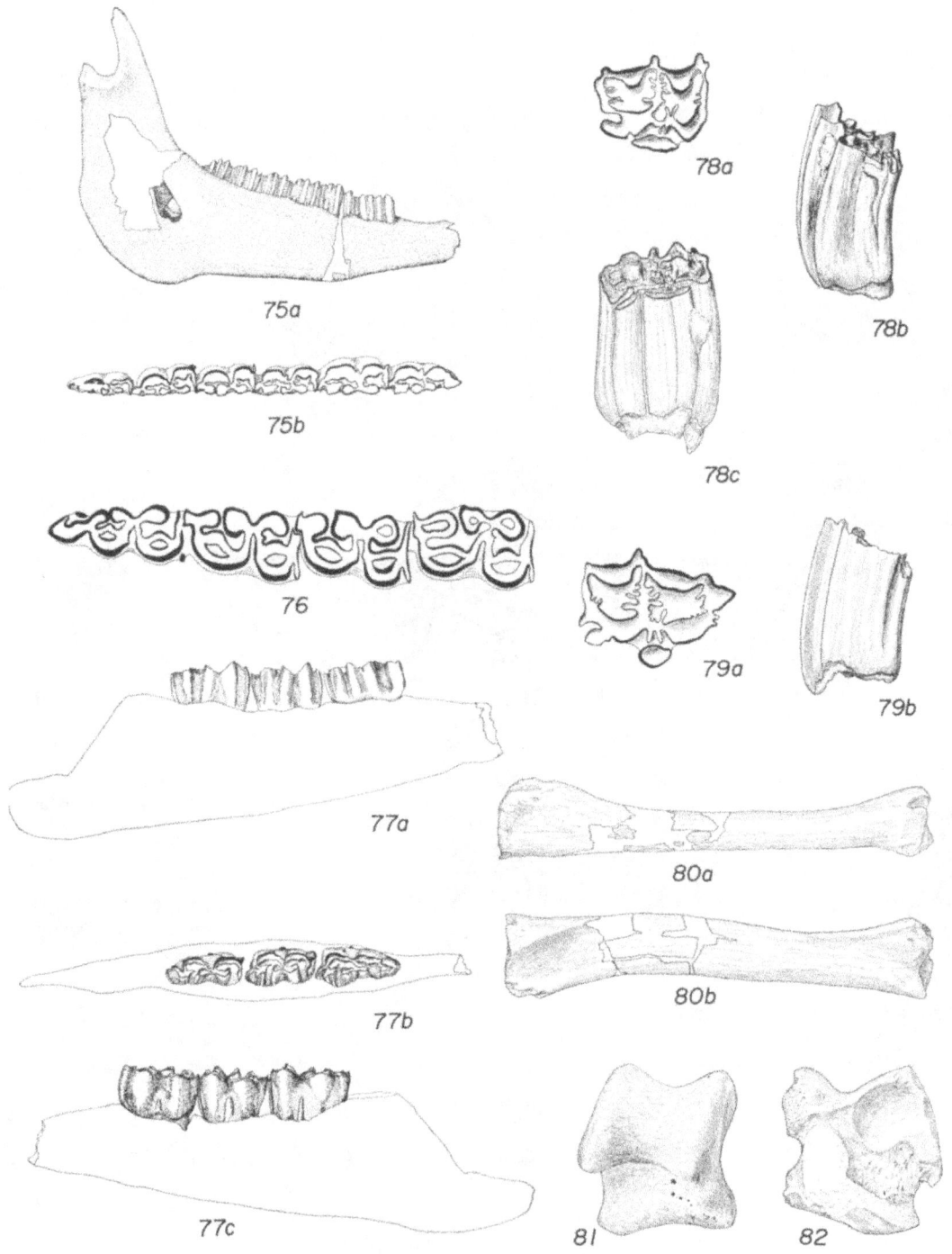

FIGS. 75–82. *Hipparion* and Tapiridae.

With the available material there is no basis for considering the Black Butte material of a different species than *H. condoni* from the Ellensberg formation of Washington.

Hipparion forcei of the Black Hawk Ranch fauna of California has a shorter hypoconal groove than the Black Butte species, the protocone connects to the protoconule in moderate wear, the exterior fossette borders are more simple, and the hypostylid is not as highly developed. *H. forcei* is about the same size as the Black Butte species. *Hipparion mohavense* of the Ricardo formation of southern California does not have a flattened lingual border on the protocone, and the protocone is rounder in most stages of wear. The fossettes are similar in their complexity in some specimens to the Black Butte species. The crown is straighter in *H. mohavense*. The protocone becomes attached to the protoconule in late wear rather commonly in *H. mohavense* whereas it apparently never does in the Black Butte species. The development of the paralophid and hypostylid is similar in the two forms compared. The Ricardo species is certainly closer to the Black Butte species than *H. forcei* but differs in a number of fundamental characteristics.

A number of North China species of *Hipparion*, discussed by Sefve (1927) have characteristics similar to the Black Butte species. Characteristics of the Black Butte species, such as well-developed protostyle and hypostyle, complicated fossette borders, extention of the postprotoconal valley into the prefossette lingually flattened protocones, strong parastylids, and squared off metastylid in lower molars, suggest closer affinities of this species with the North China species than with the California species. *Hipparion anthyoni* from Oregon also is similar to species of North China and is even more complicated in plication development than the Black Butte species. European hipparions, *H. gracile*, in particular, tend to differ from the North China and Oregon species in the same way that California species do.

FAMILY TAPIRIDAE

A left upper M2 fragment (F-5947) from locality UO 2334, navicular (F-5664) from locality UO 2333, and two astragali (F-6658) locality UO 2353 and F-5825, locality UO 2348, all represent a tapir. The astragalus is slightly larger than in *Tapirus*. The neck of the astragalus is shorter than in *Tapirus*. The tooth fragment consists of paracone, part of protoloph, and metaloph. On this fragment there is a small mesostyle giving way to small cingulum on the metacone. The material does not allow generic identification (figs. 81, 82).

FAMILY RHINOCEROTIDAE

Aphelops sp.

Three jaws, nine loose teeth, astragalus, radius, ulna, and isolated incisors are assigned to *Aphelops*.

Left lower jaw (F-10263) is complete and includes the symphysis. I-M3 are present (fig. 83). Right lower jaw (F-10528) is broken at the anterior edge of the P3. The P3-M3 are present (fig. 84). Right lower jaw (F-6695) is broken anterior to the P3. P3-M3 present. Most of the ascending ramus is missing.

Upper dentition: In M3 the crotchet is not connected. It is connected in well-worn P4's and M1's. M3 relatively unworn lower crowned than *Teleoceras*.

Lower dentition: Incisor roughly triangular. Anterior labial and anterior lingual faces slightly convex.

75. *Hipparion condoni* Merriam, left lower jaw, UOMNH F-10283, loc. 2448. Medial view ×0.25 (75a). Dorsal view ×0.25 (75b).
76. *Hipparion condoni* Merriam, right lower jaw, UOMNH F-11020, loc. 2448. Occlusal view of dentition P4-M3 ×1.
77. *Hipparion condoni* Merriam, right lower deciduous dentition UOMNH F-10285, loc. 2448. Labial view dP2-dP4 ×0.5 (77a). Occlusal view ×0.5 (77b). Lingual view ×0.5 (77c).
78. *Hipparion condoni* Merriam, upper M1, UOMNH F-10983, loc. 2448. Occlusal view ×1 (78a). Posterior view ×1 (78b). Lingual view ×1 (78c).
79. *Hipparion condoni* Merriam, upper right P2, UOMNH F-10984, loc. 2448. Occlusal view ×1 (79a). Posterior view ×1 (79b).
80. *Hipparion condoni* Merriam, tibia, UOMNH F-10796, loc. 2448. Lateral view ×0.25 (80a). Posterior view ×0.25 (80b).
81. Tapiridae, astragalus, UOMNH F-6658, loc. 2353. Anterior view ×0.5.
82. Tapiridae astragalus, UOMNH F-5825, loc. 2348. Ventral view ×0.5.

TABLE 15

MEASUREMENTS OF LOWER JAWS OF *Aphelops*

	F-10263	F-10528	F-6695
P2-M3	270	—	—
P3-M3	230	223	200
P2-AP	33.2	—	—
P3-AP	38.7	37.9	37.1
P4-AP	41.9	42.3	38.1
M1-AP	45.0	43.2	42.1
M2-AP	46.4	45.5	45.2
M3-AP	46.7	49.1	44.6
Width incisor at alveoli	26.0	—	—
Depth jaw below M3 (labial)	97.0	82.3	—
P4	82.0	78.7	Distorted
Height, angle to condyle	190 (distorted)	215	Missing
Length symphysis	103.3	—	—
Total length incisor alveoli to condyle	495	—	—

Posterior face nearly flat. Only slightly curved antero-posteriorly. P4 and M1 similar in size. M2 and M3 larger and P3 and P2 smaller. Not as high crowned as in *Teleoceras*. Depth of the jaw does not change greatly from M3-P3. Astragalus similar to living *Dicerorhinus*, not shortened as in *Teleoceras*. The facets are curved as in *Dicerorhinus*. The ginglymus valley is well defined and deep as in *Dicerorhinus*. Skeletal elements also similar (see figs. 83–87).

Order ARTIODACTYLA

Family TAYASSUIDAE

Prosthennops sp.

A single upper second molar from UO 2348, F-6007, indicates the presence of a peccary in the fauna. The tooth is rather simple without cingulum or accessory cusps. Measurements are AP-15.0 mm. T-14.8 mm. at the base of the crown. This material is inadequate for anything more than generic identification.

Family MERYCOIDODONTIDAE

Ustatochoerus sp.

An upper right canine, right maxillary fragment and several ectoloph fragments of teeth are assignable to *Ustatochoerus* (figs. 92, 93).

TABLE 16

MEASUREMENTS OF MANDIBLES OF *Procamelus* cf. *grandis* and *Megatylopus* cf. *gigas**

	F-10290	F-6222	F-10259	F-10284	F-10287*
AP diameter canine	17.6				
T diameter canine	12.1				
AP diameter P1	13.6	10.3			13.2
T diameter P1	9.1	6.8			7.1
AP diameter P2	10.6	11.0			
T diameter P2	5.7	5.9			
AP diameter P3	16.0	15.2	15.0	14.3	
T diameter P3	6.5	6.7	6.5	6.4	
AP diameter P4	20.4	18.3	18.9	18.8	26.5
T diameter P4	8.6	9.5	9.3	9.9	15.3
AP diameter M1	31.9	23.3	29.7	29.8	34.9
T diameter M1	15.5	15.6	15.9	17.5	22.0
AP diameter M2	38.3	30.3	37.5	35.5	41.4
T diameter M2	16.6	17.3	16.9	23.3	24.4
AP diameter M3	46.9	46.4	47.5	48.7	52.4
T diameter M3	16.0	17.6	15.5	18.2	19.6
P2-M3	163.0	149.0	160.0+	178	
Diastema C-P1	31.0			23+	
Diastema P1-P2	42.8	36.7	35.8	33.2	44.1
Depth of mandible below M3	72.0	67.7	62.3	66.0	69.5
Thickness of mandible below M3	23.6	25.3	26.4	28.3	34.8

Specimen numbers with asterisk are *Megatylopus*.

TABLE 17

MEASUREMENTS OF LIMB ELEMENTS OF *Procamelus* CF. *grandis* AND *Megatylopus* CF. *gigas**

(IN MILLIMETERS)

	F-10280	F-10487	F-10500*	F-10496*
Metatarsal length	480	480	500	
Diameter proximal end	64	65	74.7	72.0
Transverse diameter midshaft	42.5	44.0	50.5	51.9

	F-10513	F-10274	F-10293*	
Metacarpal length	417	419	542	
Diameter proximal end	68.7	65	82.6	
Transverse diameter at midshaft	39.8	38.7	53.5	

	F-10797	F-10495	F-10292*	
Length ulna	514	547	643	
Width at distal end	65.5	75.5	82.4	

			F-10509*	F-5937*
Length of tibia			650	650
Width of distal end			101	108

* *Megatylopus*.

Family CAMELIDAE

Procamelus cf. *grandis* Gregory (1942)

Seven lower jaws, numerous limb elements, and loose teeth are referable to *P. grandis*. The Black Butte material is slightly larger than the typical material from Big Springs Canyon, South Dakota, but agrees in most other characteristics. The premolars are blade-like, the lower first premolar is apparently single rooted, and the diastema are long. The second premolar is present. The Black Butte material differs in its greater size and longer metapodials and cervical vertebrae.

This is the most common mammal in the fauna. It is best known from quarry 3 and quarry 2. However, almost every other locality has material referable to this camel. Measurements of typical skeletal elements and illustrations of them are included.

83. *Aphelops* sp., left lower jaw, UOMNH F-10263, loc. 2448. Lateral view of jaw and occlusal view of dentition P2-M3 ×0.25.
84. *Aphelops* sp., right lower jaw, UOMNH F-10528, loc. 2448. Lateral view of jaw and occlusal view dentition P3-M3, ×0.25.
85. *Aphelops* sp., upper ?P4, UOMNH F-10346, loc. 2448. Occlusal view ×0.5.
86. *Aphelops* sp., upper M3, UOMNH F-10342, loc. 2448. Occlusal view ×0.5.
87. *Aphelops* sp., upper ?P4, UOMNH, F-11012, loc. 2448, Occlusal view ×0.5.

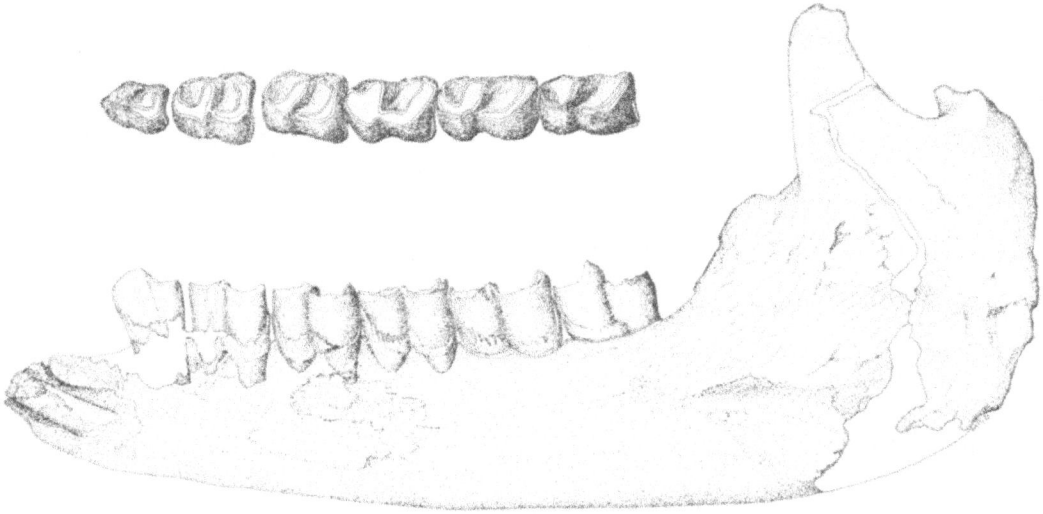

83

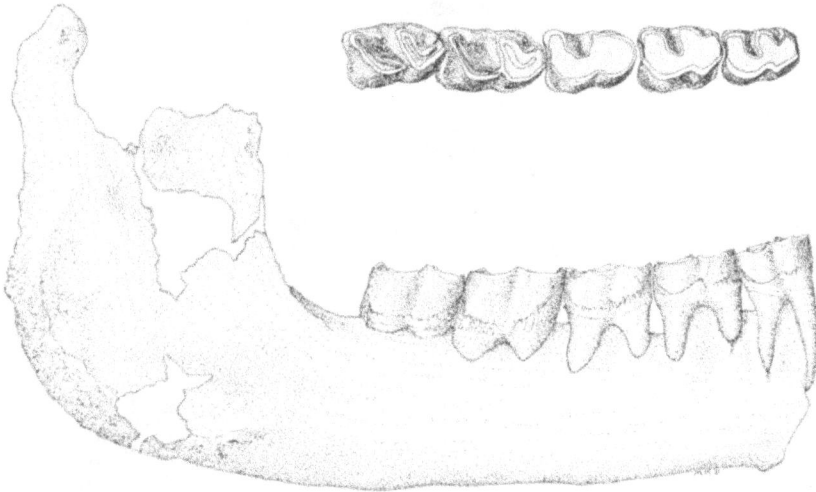

84

85

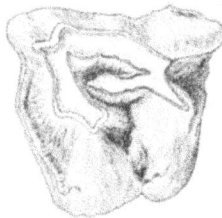

86

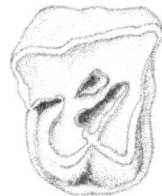

87

FIGS. 83–87. *Aphelops.*

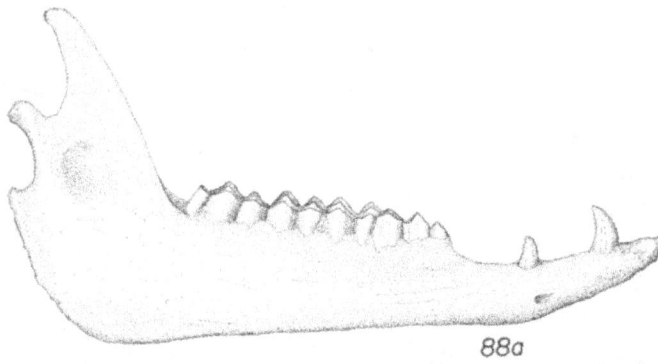

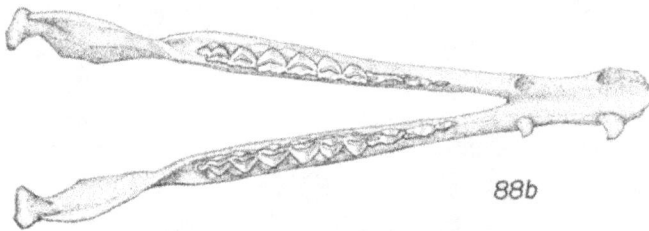

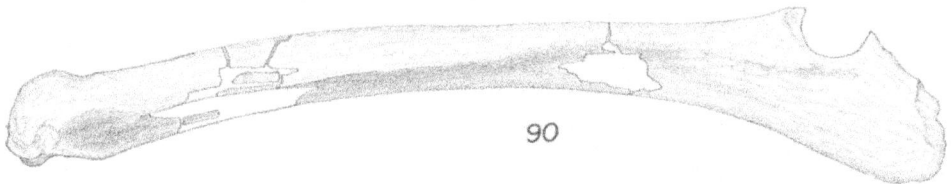

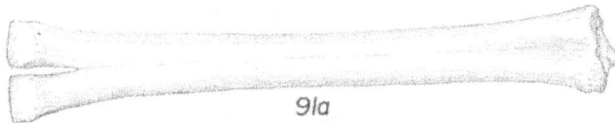

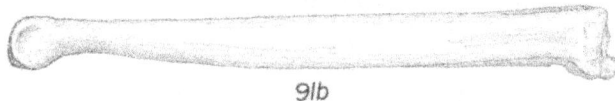

FIGS. 88–93. Artiodactyla of the Black Butte fauna.

A number of smaller metapodials are present in the collection. Most of these are immature specimens suggesting that they are also referable to *P. grandis*. Other skeletal elements including loose teeth do not suggest that there is a smaller camel present in the fauna (figs. 88–91).

Megatylopus cf. *gigas* Matthew and Cook

A lower jaw with P4-M3 present and alveoli of P1 and P3 from quarry 3 represents *Megatylopus*. A number of metapodials, loose teeth, calcani, astragali, and other skeletal elements are also assignable to this genus on the basis of size and proportions. This camel is also best known from quarry 3. It occurs quite often in other localities. Its large size makes it relatively easy to recognize. It, however, never occurs in great abundance as does *Procamelus*. Typical elements are figured (figs. 94–98).

REFERENCES

BARBOUR, E. H. 1927. Preliminary notice of a new proboscidean, *Amebelodon fricki* gen. et sp. nov. *Bull. Neb. State Museum* 1: 131–134.

—— 1929. *Torynobelodon loomisi*, gen. et sp. nov. *Bull. Neb. State Museum* 1: 147–153.

—— 1931. A new amebelodont, *Torynobelodon barnumbrowni*, sp. nov. A preliminary report. *Bull. Neb. State Museum* 1: 191–198.

—— 1932. The mandible of *Torynobelodon barnumbrowni*. *Amer. Jour. Sci.* 5 (24): 214–220.

BORISSIAK, A. A. 1927. On a new mastodon from the Chakrak beds (Middle Miocene) of the Kuban region, *Platybelodon donovi*, n. gen. n. sp. *Ann. Soc. Pal. Russia* 7: 105–120.

BRYANT, M. D. 1945. Phylogeny of the Nearctic Sciuridae. *Amer. Mid. Nat.* 33: 357–390.

COPE, E. D. 1890. On two new species of Mustelidae from the Loup Fork Miocene of Nebraska. *Amer. Nat.* 24: 950–952.

—— 1892. A hyena and other Carnivora from Texas. *Amer. Nat.* 26: 1028–1029.

—— 1893. A preliminary report on the vertebrate paleontology of the Llano Estacado. *Fourth Ann. Rept. Geol. Surv. Texas*, 1–136.

FURLONG, E. L. 1932. A new genus of otter from the Pliocene of the Northern Great Basin Province. *Carn. Inst. Wash. Publ.* 418: 93–103.

GAZIN, C. L. 1933. A Miocene mammalian fauna from southeastern Oregon. *Carn. Inst. Wash. Publ.* 418: 37–86.

—— 1934. Upper Pliocene mustelids from the Snake River Basin of Idaho. *Jour. Mammal.* 15 (2): 137–149.

—— 1937. Notes on fossil mustelids from the Upper Pliocene of Idaho and Texas. *Jour. Mammal.* 18: 363–364.

GREGORY, J. T. 1942. Pliocene vertebrates from Big Spring Canyon, South Dakota. *Univ. Calif. Publ. Dept. Geol. Sci.* 26: 307–446.

—— 1945. An Amebelodon jaw from the Texas panhandle. *Univ. Texas Publ.* 4401: 477–484.

HALL, E. R. 1930*a*. Rodents and lagomorphs from the Late Tertiary of Fish Lake Valley, Nevada. *Univ. Cal. Pub. Bull. Dept. Geol. Sci.* 19: 295–312.

—— 1930*b*. A bassarisk and a new mustelid from the later Tertiary of California. *Jour. Mammal.* 11: 23–26.

—— 1935. A new mustelid genus from the Pliocene of California. *Jour. Mammal.* 16: 137–138.

—— 1944. A new genus of American Pliocene badger, with remarks on the relationships of badgers of the Northern Hemisphere. *Carn. Inst. Wash. Publ.* 551: 9–23.

HIBBARD, C. W. 1957. Notes on late Cenozoic shrews. *Trans. Kansas Acad. Sci.* 60 (4): 327–336.

HIBBARD, C. W., and LESTER F. PHILLIS. 1945. The occurrence of Eucastor and Epigaulus in the Lower Pliocene of Trego County, Kansas. *Univ. Kansas. Sci. Bull.* 30 (2): 549–555.

KORMOS, T. 1931. *Pannonictis pliocaenica* n.g., n. sp., a new giant mustelid from the late Pliocene of Hungary. *Evkonkir. Fold. Intezet.* 29: 167–177.

MACDONALD, J. R. 1956. A new Clarendonian mammalian fauna from the Truckee formation of western Nevada. *Jour. Paleo.* 30 (1): 186–202.

MATTHEW, W. D. 1924. Third contribution to the Snake Creek faunas. *Bull. Amer. Mus. Nat. Hist.* 50: 59–210.

—— 1932. New fossil mammals from the Snake Creek quarries. *Amer. Mus. Novit.* 540: 1–8.

MATTHEW, W. D., and HAROLD COOK. 1909. A Pliocene fauna from western Nebraska. *Bull. Amer. Mus. Nat. Hist.* 26: 361–414.

MERRIAM, J. C. 1915. New species of the Hipparion group from the Pacific Coast and Great Basin provinces of North America. *Univ. Cal. Pub. Bull. Dept. Geol.* 9: 1–8.

—— 1919. Tertiary mammalian faunas of the Mohave Desert. *Univ. Calif. Publ. Bull. Dept. Geol. Sci.* 11: 437–585.

MERRIAM, J. C., C. STOCK, and C. L. MOODY. 1925. The Pliocene rattlesnake formation and fauna of Eastern Oregon, with notes on the geology of the rattlesnake and Mascall deposits. *Carn. Inst. Wash. Pub.* 347: 43–92.

OSBORN, H. F. 1929. The revival of Central Asiatic life. *Nat. Hist.* 29: 3–16.

—— 1933. *Serbelodon burnhami*, a new shovel tusker from California. *Amer. Mus. Novit.* 639: 105.

---

88. *Procamelus* cf. *grandis* Gregory, mandible, UOMNH F-10290, loc. 2448, view right side ×0.25 (88*a*). Dorsal view ×0.25 (88*b*).

89. *Procamelus* cf. *grandis* Gregory, lower right jaw UOMNH F-10290, loc. 2448. Occlusal view dentition ×0.5.

90. *Procamelus* cf. *grandis* Gregory, radi-ulna UOMNH F-10797, loc. 2448. ×0.25.

91. *Procamelus* cf. *grandis* Gregory, metatarsal, UOMNH F-10795, loc. 2448. Dorsal view ×0.25 (91*a*). Lateral view ×0.25 (91*b*).

92. ?*Ustatochoerus* sp. incisor, UOMNH F-5756, loc. 2341. Lateral view ×1.5 (92*a*). Posterior view ×1.5 (92*b*).

93. ?*Ustatochoerus* sp. incisor, UOMNH F-5796, loc. 2340. Lateral view ×1.5 (93*a*). Posterior view (93*b*).

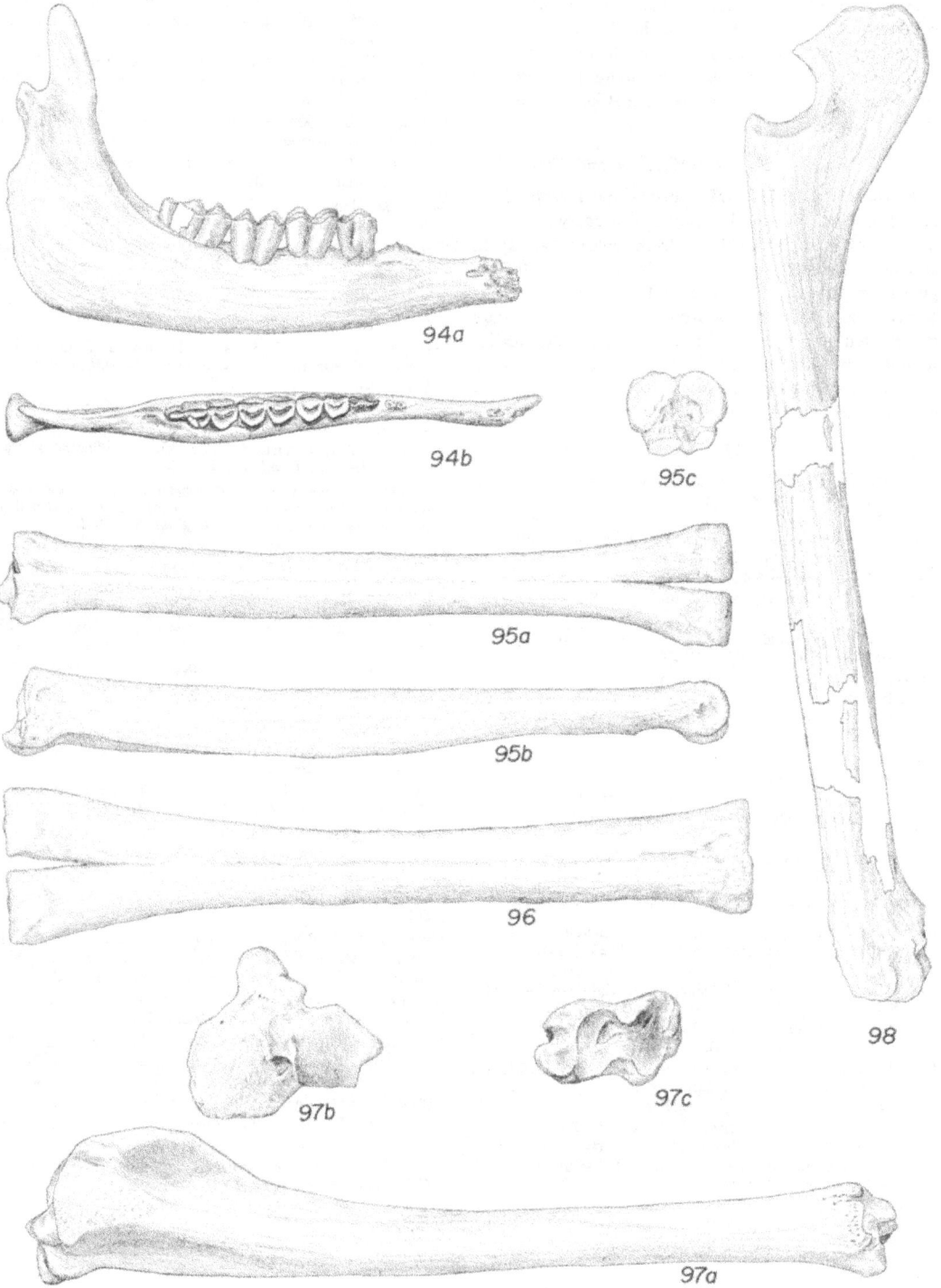

FIGS. 94–98. *Megatylopus.*

RICHEY, K. A. 1938. A new dog from the Black Hawk Ranch fauna, Mt. Diablo, California. *Univ. Calif. Publ. Dept. Geol. Sci.* **24**: 303–308.

—— 1948. Lower Pliocene horses from Black Hawk Ranch, Mount Diablo, California. *Univ. Calif. Publ. Bull. Dept. Geol. Sci.* **28**: 1–44.

94. *Megatylopus* cf. *gigas* Matthew and Cook, right lower jaw UOMNH F-10287, loc. 2448. Lateral view ×0.25 (94*a*). Dorsal view ×0.25 (94*b*).
95. *Megatylopus* cf. *gigas* Matthew and Cook, metatarsal UOMNH F-10794, loc. 2448. Dorsal view ×0.25 (95*a*). Lateral view ×0.25 (95*b*). View proximal end ×0.25 (95*c*).
96. *Megatylopus* cf. *gigas* Matthew and Cook, metacarpal UOMNH F-10293, loc. 2448. Dorsal view ×0.25.
97. *Megatylopus* cf. *gigas* Matthew and Cook, tibia UOMNH F-5937, loc. 2340. Lateral view ×0.25 (97*a*). View proximal end ×0.25 (97*b*). View distal end ×0.25 (97*c*).
98. *Megatylopus* cf. *gigas* Matthew and Cook, radi-ulna UOMNH F-10292, loc. 2448. Lateral view ×0.25.

RUSSELL, D. E. 1956. Clarendonian fauna of Juntura, Oregon. Unpublished thesis, University of California, Berkeley.

SAVAGE, D. E. 1955. Nonmarine lower Pliocene sediments in California. *Univ. Calif. Publ. Geol. Sci.* **31** (1): 1–26.

SEFVE, I. 1927. Die Hipparionen Nord-chinas. *Pal. Sinica* **4** (2): 1–95.

SHOTWELL, J. A. 1955. Review of the Pliocene beaver *Dipoides*. *Jour. Paleo.* **29**: 129–144.

—— 1958. Evolution and biogeography of the aplodontid and mylagaulid rodents. *Evolution* **12** (4): 451–484.

STIRTON, R. A. 1935. A review of the Tertiary beavers. *Univ. Calif. Publ. Bull. Dept. Geol. Sci.* **23**: 391–458.

VANDERHOOF, V. L. 1937. A study of the Miocene Sirenian *Desmostylus*. *Univ. Calif. Publ. Bull. Dept. Geol. Sci.* **24**: 169–262.

ZDANSKY, O. 1924. Jungtertiare Carnivoren Chinas. *Pal. Sinica* (C) **2** (1): 1–155.

—— 1927. Weitere Bemerkungen uber Fossile Carnivoren aus China. *Pal. Sinica* (C) **4** fasc. 4: 1–28.

5. MAMMALIAN FAUNA OF THE DREWSEY FORMATION, BARTLETT MOUNTAIN, DRINKWATER AND OTIS BASIN LOCAL FAUNAS

J. ARNOLD SHOTWELL

Museum of Natural History, University of Oregon

OCCURRENCE

The Drewsey formation is for the most part distributed in the western portion of the Juntura Basin. Some exposures do occur in the eastern edge and have produced fossils, UO loc. 2360 in particular. Most localities, however, are to the west of Drinkwater Pass. These are in three different areas and take their names from their geographic occurrence. A number of localities are found on the west slope of Drinkwater Pass. These are grouped as the Drinkwater local fauna. They occur in a heavily indurated brown tuffaceous sandstone which is well exposed locally. In Otis Basin a single productive locality is known. It is the basis for the Otis Basin fauna from UO loc. 2347. On the north flank of Bartlett Mountain a number of localities are grouped as the basis for the Bartlett Mountain fauna. This name has appeared formerly in the literature (Wilson, 1937b) with some later confusion as to its age. I believe from studying the materials originally assigned to this fauna that specimens from both the Drewsey and Juntura formations were inadvertently mixed in the earlier collections. This may be due to the origin of the Drewsey from former Juntura formation sediments. The Juntura formation is exposed at Sitz Hot Spring and Tom Delahney Butte to the northwest. The name, Bartlett Mountain local fauna, is still the obvious name for the adjacent fauna. Bartlett Mountain is the most prominent feature locally. Changing the name would be only confusing. The name Stinkingwater from nearby Stinkingwater Creek is another possibility but has been previously applied to an older flora by Chaney (1959).

Fossil mammals have been known from this local area for a long time. Many of the early collections of this area are part of the California Institute of Technology collection now at the Los Angeles County Museum. Their localities are noted on the map where possible or their identity with subsequent localities is noted.

The first known rodents of the Bartlett Mountain local fauna were cited by Wilson (1937b: 33). Later Hall (1944) described a badger collected by Al Brown of Burns, Oregon, probably from the same locality. Mr. Brown's badger came from the locality we have called UO loc. 2239. This and other nearby localities probably equal CIT loc. 107. Shotwell (1958) has previously made reference to the mylagaulid and aplodontid rodents from the Bartlett Mountain and Otis Basin local faunas.

This chapter will be concerned only with the age and systematics of the local faunas. The ecological aspects of the faunas are considered in chapter 1.

AGE AND CORRELATION

The joint occurrence of *Dipoides, Hipparion, Pliohippus, Sphenophalos, Mylagaulus, Liodontia, Pliosaccomys, Amebelodon, Prosthennops,* and *Teleoceras* indicates the Hemphillian age of this fauna in the North American Continental Tertiary sequence. The Bartlett Mountain, Otis Basin, and Drinkwater local faunas are similar in age to the Rome, Rattlesnake, and McKay Reservoir local faunas of Oregon and the Thousand Creek local fauna of Nevada. Several poorly known faunas to the east of the Juntura Basin near Harper and Vale in Oregon are also probably equivalent in age. Several species present in the local faunas of the Drewsey formation are known from the Rome local fauna. They are *Pliohippus spectans, Microtoscoptes disjunctus,* and *Dipoides stirtoni.* Some important species in the local faunas of the Drewsey formation are not known at Rome. These are *Hipparion condoni* (no hipparion horses are known from Rome), *Liodontia furlongi,* and *Hystricops. Hystricops* is a rare form and this may account for its absence in the Rome collection. The absence of *Liodontia furlongi* and the hipparion horses at Rome is significant. This probably indicates a difference in environments represented rather than age. Although the collection from the Rattlesnake local fauna of the John Day Basin is scrappy, hipparion horses and *Liodontia furlongi* are present. *Pliohippus spectans* and *Dipoides stirtoni* are also present in the Rattlesnake fauna. The McKay Reservoir local fauna, to the north along the Columbia River in Oregon, contains a different species of *Dipoides, Hipparion* of unknown species, *Liodontia furlongi* and *Pliosaccomys dubius.* The McKay fauna does not include *Pliohippus. Pliohippus* apparently did not reach this region (Shotwell, 1961).

COMPOSITE LIST OF LOCAL FAUNAS OF THE DREWSEY FORMATION

Order Lagomorpha
  Family Leporidae
    Hypolagus cf. *oregonensis* Shotwell, 1956
Order Rodentia
  Family Aplodontidae
    Liodontia furlongi Gazin, 1932
  Family Mylagaulidae
    Mylagaulus sp.

Family Sciuridae
  Citellus sp.
Family Castoridae
  Hystricops browni n. sp.
  Dipoides stirtoni Wilson, 1934
Family Geomyidae
  Pliosaccomys dubius Wilson, 1936
Family Cricetidae
  Microtoscoptes disjunctus (Wilson, 1937*a*)
Order Carnivora
  Family Canidae
    Osteoborus cf. *cyonoides* (Martin, 1928)
  Family Mustelidae
    Pliotaxidea nevadensis (Butterworth, 1916)
Order Proboscidea
  Family Gomphotheriidae
    Amebelodon? sp.
  Family Mammutidae
    Mammut sp.
Order Perissodactyla
  Family Equidae
    Hipparion cf. *condoni* Merriam, 1915
    Pliohippus spectans (Cope, 1880)
  Family Rhinocerotidae
    Teleoceras sp.
Order Artiodactyla
  Family Tayassuidae
    Prosthennops sp.
  Family Camelidae
    Procamelus sp.
    Megatylopus sp.
  Family Antilocapridae
    Sphenophalos sp.

SYSTEMATICS

Order LAGOMORPHA

Hypolagus cf. *oregonensis* Shotwell, 1956

Rabbits are among the most common small mammals represented in the collections. Numerous isolated teeth, jaw fragments, and skeletal element fragments are known. The upper molars have serrate reentrant borders. In the upper P2 there are two reentrant grooves on the anterior face. One is much deeper than the other. Both have cement-filled stria the full length of the tooth. Lower P3 often has two shallow cement-filled grooves. The labial reentrant has serrated border. Just lingual to the end of the reentrant is a small pit. I cannot determine whether this pit is surrounded with enamel or not. These characters of the dentition are similar to many species of *Hypolagus* but probably most closely approach *H. oregonensis* Shotwell (1956).

Order RODENTIA

Family APLODONTIDAE

Liodontia furlongi Gazin, 1932

This aplodontid rodent is common in the collections. About twenty isolated teeth, a nearly complete lower jaw, and two fragmental skulls are present. These were reported earlier by Shotwell (1958).

Family MYLAGAULIDAE

Mylagaulus sp.

Our collections consist of a number of loose teeth and a nearly complete lower jaw. A complete skull is in the CIT collections. These have been illustrated and described in detail in Shotwell (1958). Additional details of the important upper and lower fourth premolars are as follows:

Moderately worn lower premolars have 6–7 lakes depending on whether the antifossettid is present or not. The six basic mylagaulid lakes are present. They are lengthened and tend to be in pairs antero-posteriorly. Hypostriid and sometimes the protostriid are present as shallow grooves. In heavy wear the lower fourth premolars display nine lakes. The three additional lakes are derived from fragmentation of para, meta, and mesofossettid. These teeth are smaller than most specimens of other Hemphillian *Mylagaulus* examined and slightly larger than the Clarendonian Black Butte *Mylagaulus*. Characters of moderately worn teeth are similar to Clarendonian species in simplicity, but little-worn specimens display the typical fragmented pattern seen in Hemphillian *Mylagaulus*.

TABLE 18

MEASUREMENTS OF LOWER FOURTH PREMOLARS OF
Mylagaulus

	AP	T	Comments
F-6113	11.4 mm.	4.8 mm.	Hypostriid a shallow groove
F-6115	12.1 mm.	5.0 mm.	Hypostriid a shallow groove
F-15697	11.4 mm.	4.7 mm.	Hypostriid a shallow groove
F-15673	11.7 mm.	5.6 mm.	Hypostriid and protostriid present

A single upper fourth premolar is known. It is nearly identical with upper premolars known from the Rome, Oregon, fauna (Shotwell, 1958: fig. 15*B*).

These specimens indicate that the species of *Mylagaulus* in this fauna is transitional between those of the Clarendonian Black Butte fauna and the Hemphillian Rome fauna in size and some occlusal pattern characteristics of the teeth.

Family SCIURIDAE

Citellus sp.

A single upper right fourth premolar and an incisor from UO loc. 2347 indicate the presence of a small

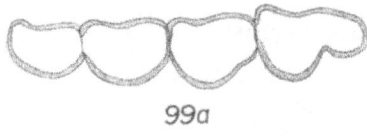

99a

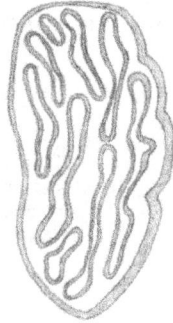

100

101

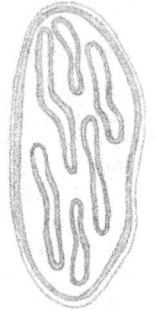

102

99b

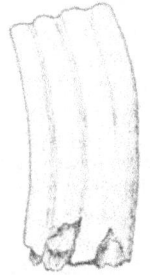

106a

106b

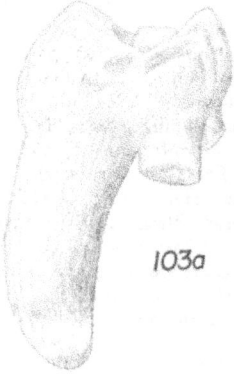

103a

103b

107

110

104a

104b

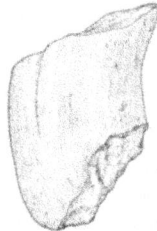

104c

108

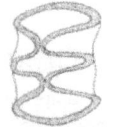

111a

105a

105b

105c

109

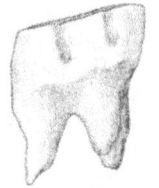

111b

Figs. 99–111. Rodentia of the Drewsey formation.

ground squirrel in the fauna. The parastyle of the premolar is well developed. The anterior cingulum attaches low on the protocone. The metaloph does not connect to the protocone. The metacone is bulbous. The protoloph connects through the protocone to the posterior cingulum to form a U. The tooth is rather low-crowned but little worn. Specific assignment is not possible but the characters in the single premolar known suggest a small otospermophilid ground squirrel.

FAMILY CRICETIDAE

Microtoscoptes disjunctus (Wilson, 1937a)

A single lower left second molar of this aberrant microtine was collected from UO loc. 2239 (see fig. 111).

FAMILY GEOMYIDAE

Pliosaccomys dubius Wilson, 1936

A single upper fourth premolar of this gopher is known from UO loc. 2347.

FAMILY CASTORIDAE

Dipoides cf. stirtoni Wilson, 1934

Loose teeth and fragmental skeletal elements of a small beaver are common in several of the localities worked. They were most common in UO loc. 2356. Most of the following comments are based on the specimens from this locality.

99. *Liodontia furlongi* Gazin, right lower jaw, UOMNH F-6101, loc. 2347. Occlusal view dentition ×5 (99a). Lateral view of jaw ×2.5 (99b).
100. *Mylagaulus* sp. upper P4, UOMNH F-15693, loc. 2358. Occlusal view ×4.
101. *Mylagaulus* sp. upper P4, UOMNH F-6115, loc. 2347. Occlusal view ×4.
102. *Mylagaulus* sp. lower P4, UOMNH F-15691, loc. 2356. Occlusal view ×4.
103. *Citellus* sp. upper P4, UOMNH F-15688, loc. 2347. Posterior view ×15 (103a). Occlusal view ×15 (103b).
104. *Hystricops browni* Shotwell, n. sp. Holotype UOMNH F-15696, loc. 2239. Occlusal view P4 ×2.5 (104a). Lateral views P4 ×2.5 (104b, 104c).
105. *Hystricops browni* Shotwell, n. sp. UOMNH F-15696, loc. 2239, occlusal view M1 ×2.5 (105a). Lateral views ×2.5 (105b, 105c).
106. *Dipoides* cf. *stirtoni* Wilson, upper M3, UOMNH F-9555, loc. 2358. Occlusal view ×4 (106a). Lateral view ×4 (106b).
107. *Dipoides* cf. *stirtoni* Wilson, upper M3, UOMNH F-9557, loc. 2358. Occlusal view ×4.
108. *Dipoides* cf. *stirtoni* Wilson, upper M3, UOMNH F-9566, loc. 2358. Occlusal view ×4.
109. *Dipoides* cf. *stirtoni* Wilson, upper M3, UOMNH F-9563, loc. 2358. Occlusal view ×4.
110. *Pliosaccomys dubius* Wilson, upper P4, UOMNH F-15687, loc. 2347. Occlusal view ×10.
111. *Microtoscoptes disjunctus* (Wilson), lower left M2, UOMNH F-15689, loc. 2239. Occlusal view ×10 (111a). Lateral view ×10 (111b).

All lower fourth premolars collected display a full-length parastriid as in *D. stirtoni*. One specimen has closed fossettids at the base which is rare in *Dipoides* but occasionally occurs in *D. stirtoni*. In the upper fourth premolar the parastria is complete in the specimens collected. The anterior loph is conected to the median loph by a narrow isthmus. The hypoflexus and paraflexus overlap. A good wear series of the upper third molar is represented in the collection. Young individuals display a long but incomplete parastria, full-length meso- and metastria. The parastria wears to a parafossette which is lost in still later wear. The mesoflexus and metaflexus are connected for the full length of the tooth isolating the mesostyle as an enamel tube (fig. 107–109). The only known probable third molar in the topotypic *D. stirtoni* material from Rome has a meta- and mesostria but they are not connected to isolate the mesostyle as in the present material. This is possibly of little significance. The upper third molar in beavers often shows many individual variants in occlusal pattern (Shotwell, 1955).

The specimens from the Drewsey formation represent a beaver in the same size range as *D. stirtoni*. The similarity of characters of the upper and lower fourth premolars further indicates that this species is probably assignable to *Dipoides stirtoni* Wilson, 1934.

Hystricops browni n. sp.

Holotype: An associated partial upper dentition including P4-M1 on each side. UOMNH F-15696.

Type locality: UO loc. 2239. Drewsey formation, Harney County, Oregon.

Diagnosis: Similar in size to *Hystriocops* sp. of Black Butte, but with more persistent stria and striids.

Named in appreciation to Al Brown of Burns, Oregon. Mr. Brown has been of help to many paleontologists for many years and has been an enthusiastic proponent of important features of earth history present in Harney County.

Description: The four upper teeth which make up the type of the new species were found closely associated although not in place. They consist of opposite upper fourth premolars and first molars. The identical wear of opposing teeth and their occurrence indicate that they are probably from a single individual. On the fourth premolar the hypostria extends to within 4.5 mm. of the base of the enamel. The parastria extends to within 3.2 mm. of the base of the enamel. The metastria is closed at the stage of wear seen in the specimens and is represented by two inequally sized fossettes. The paraflexus and hypoflexus bypass each other. The mesoflexus bends sharply posteriorly near its distal end. See fig. 104.

The first molar is a much smaller tooth than the premolar. The hypostria is the only open stria. There is a parafossette and a metafossette at the stage of wear of the specimens. Roots are poorly developed but present and are similar to those of *Eucastor* molars.

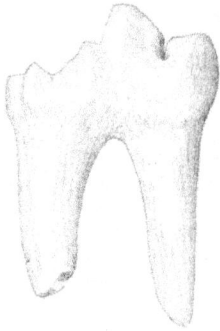

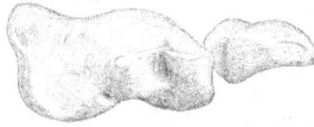

112b

118

112a

114

113

115

119a

116a

116b

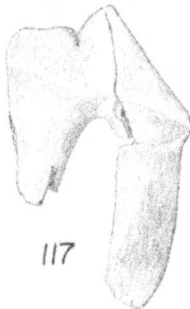

117

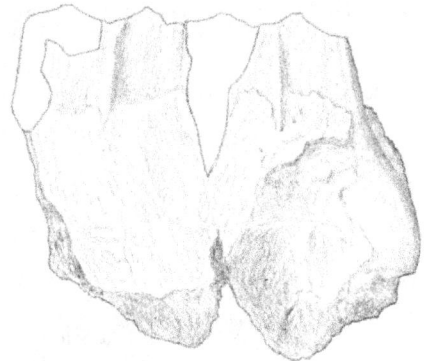

119b

FIGS. 112–119. Larger mammals of the Drewsey formation.

The paraflexus and hypoflexus meet squarely. The mesofossette is crescentic.

Comparisons: This species is very similar to the Black Butte fauna species. It differs in more persistent stria; the hypostria and mesostria are a little longer than in the older species. The parastria is much longer extending to within 3.2 mm. of the base of the enamel whereas it extends to within 9.2 mm. of the base of the enamel in the Black Butte species. The worn tooth of the new species suggests that the metastria has only been closed a short time. This would indicate a much more persistent metastria than in the Black Butte species.

The upper fourth premolars of the two species are similar in size and at different stages of wear probably have similar occlusal patterns. The phylogenetic step between the two species is a common one in beavers. They are undoubtedly directly related.

TABLE 19

MEASUREMENTS UPPER TEETH *Hystricops browni*
(IN MILLIMETERS)

	AP	T	Greatest length of enamel
P4	9.6	8.7	11.3
M1	6.5	7.0	

Since no incisor material is known of either species, the character of the enamel is not known. As mentioned in the discussion of the Black Butte species, these two species strongly suggest a parallel line to the *Monosaulax-Eucastor-Dipoides* line of castoroid beavers but delayed in the development of hypsodont teeth. The Hemphillian form described here has progressed to about the stage of *Monosoulax* in the development of hypsodont teeth and associated characteristics of tooth shape and stria extension. *Hystricops,* however, is a much larger beaver than *Monosaulax.* These two lines may possibly have a common ancestry in *Monosaulax* but more probably in a still earlier form.

ORDER CARNIVORA
FAMILY CANIDAE
Osteoborus cf. *cyonoides* (Martin, 1928)

Parts of both lower jaws of a juvenile dog were found at UO loc. 2360. In association with these was an upper left DP4 presumably of the same individual. The lower jaws which are in a fragmental state have both deciduous and permanent teeth present. These include right DP3 and DP4; left DP3 and DP4. Permanent unerupted teeth include the right lateral incisor, canine, P4, M1, M2, and left M1. The unerupted permanent teeth compare closely with those of *Osteoborus cyonoides* from the Coffee Ranch locality of the Hemphill local fauna. The deciduous DP3's compare well with one illustrated by Matthew and Stirton (1930: pl. 31). There is nothing in the available material to suggest that this is not the same species. It is not surprising to find a carnivore with such a wide geographic range.

FAMILY MUSTELIDAE
Pliotaxidea nevadensis (Butterworth, 1916)

Hall (1944) described a badger skull from UO loc. 2239 and assigned it to the species indicated above. This specimen was collected by Al Brown. He confirmed the occurrence.

ORDER PROBOSCIDEA
FAMILY GOMPHOTHERIIDAE
?*Amebelodon* sp.

Several partial lower third molars and a tusk fragment from localities UO 2355 and 2357 appear to represent *Amebelodon.* Nearly every locality in the Drewsey formation has produced some mastodont tooth fragments but are not identifiable.

ORDER PERISSODACTYLA
FAMILY EQUIDAE
Pliohippus cf. *spectans* (Cope, 1880)

Several complete and fragmental teeth and limb elements represent a large *Pliohippus* in the fauna. The protocone on the upper premolars is ovoid. The hypoconal groove is short and, when wear proceeds beyond it, a small lake is present. There are several flexures in the fossette borders but they are not complicated. There is a single pli caballin fold in all the teeth seen, but they all may be premolars. The styles are sharp. The most complete tooth measures 29.6 mm. antero-posteriorly and 29.7 mm. transversely. The tooth is 54.9 mm. high from the base of the styles. The hypoconal groove is 12 mm. long from the occlusal surface of the tooth. Although this is a very large

112. *Osteoborus* cf. *cyonoides* (Martin), lower dP3-dP4, UOMNH F-18175, loc. 2360. Lateral view dP4 ×1.5 (112*a*). Occlusal view DP3-dP4 ×1.5 (112*b*).
113. *Osteoborus* cf. *cyonoides* (Martin) lower unerupted P3, UOMNH F-18175, loc. 2360. Lateral view ×1.5.
114. *Osteoborus* cf. *cyonoides* (Martin) lower dP3 UOMNH F-18175, loc. 2360. Lateral view ×1.5.
115. *Osteoborus* cf. *cyonoides* (Martin) lower unerupted M1, UOMNH F-18175, loc. 2360. Lateral view ×1.5.
116. *Osteoborus* cf. *cyonoides* (Martin) lower unerupted M2, UOMNH F-18175, loc. 2360. Occlusal view ×1.5 (116*a*). Lateral view ×1.5 (116*b*).
117. *Osteoborus* cf. *cyonoides* (Martin) upper dP4, UOMNH F-18175, loc. 2360. Lateral view ×1.5.
118. *Hipparion* cf. *condoni* Merriam, upper molar, UOMNH F-9095, loc. 2355. Occlusal view ×1.5.
119. *Sphenophalos* sp. M2-M3, UOMNH F-18304, loc. 2356. Occlusal view ×1.5 (119*a*). Lateral view ×15 (119*b*).

Pliohippus it does not exceed some specimens of *P. spectans* known from the Rattlesnake fauna of Oregon. Its characteristics are not closely comparable to the typical *P. spectans* but are very much like some of the variants from the topotypic locality of that species.

Hipparion cf. condoni Merriam, 1915

Several upper molars and premolars and metapodial fragments of a hipparion horse are known from various localities. On the upper teeth there is no plicabela. The protocone is flattened lingually giving it a rounded triangle occlusal outline. The fossette borders are similar to those of *H. condoni* from the Black Butte fauna. More material might indicate that an assignment to *H. condoni* is not justified. However, specimens from Ellensberg, Washington, The Dalles, Oregon, and Black Butte include examples close to the known material.

FAMILY RHINOCEROTIDAE

Teleoceras sp.

Rhinoceras limb elements and tooth fragments are common in the collections. A palate and partial skull of a juvenile from UO loc. 2239 confirms the opinion that this rhino is *Teleoceras*. In the skull the DP4's are still in place and the third molars are not yet erupted. The specimens compare closely with those from the Hemphillian McKay Reservoir fauna of Oregon.

ORDER ARTIODACTYLA

FAMILY TAYASSUIDAE

?Prosthennops sp.

A symphysis of a lower jaw with a fragment of the tusk in place, an ulna, metapodial fragments, a phalanx, and an astragalus are all assignable to a peccary. The species cannot be determined. The tusk is comparable with *Prosthennops*.

FAMILY ANTILOCAPRIDAE

Sphenophalos cf. nevadanus Merriam, 1911

A number of isolated teeth from localities 2356 and 2360 are comparable to teeth assigned to *Sphenophalos nevadanus* by Merriam (1911). Also present are an astragalus, and a navicular-cuboid of an antelope. No horn core material is as yet known from these localities.

FAMILY CAMELIDAE

?Megatylopus sp.

Fragments of elements of a very large camel compare with those of *Megatylopus* from the Black Butte fauna in size and other characteristics. The dental elements are too fragmentary to make a positive identification.

?Procamelus sp.

The most common camel present is a moderate-sized one similar in size to *Procamelus* from the Black Butte

fauna. Although much more common than the *?Megatylopus* in the collections from the Drewsey formation, no definite assignment can be made as to species.

LOCAL FAUNA LISTS

Bartlett Mountain local fauna
  UO localities 2355–2358, 2339, CIT 107
    Hypolagus cf. *oregonensis*
    Liodontia furlongi
    Mylagaulus sp.
    Hystricops browni
    Dipoides cf. *stirtoni*
    Microtoscoptes disjunctus
    Pliotaxidea nevadensis
    ?Amebelodon sp.
    Mammut sp.
    Hipparion cf. *condoni*
    Pliohippus spectans
    Teleoceras sp.
    Procamelus sp.
    Megatylopus sp.
    Sphenophalos sp.
Drinkwater local fauna
  UO localities 2361, 2362, 2366
    Large carnivore
    Teleoceras sp.
    Procamelus sp.
    Megatylopus sp.
Otis Basin local fauna
  UO locality 2347
    Hypolagus cf. *oregonensis*
    Liodontia furlongi
    Mylagaulus sp.
    Citellus sp.
    Pliosaccomys dubius
    Mammut sp.
    Hipparion sp.
    Pliohippus sp.
    Rhino
    Prosthennops sp.
    Procamelus sp.

REFERENCES

BUTTERWORTH, E. M. 1916. A new mustelid from the Thousand Creek Pliocene of Nevada. *Univ. Cal. Bull. Dept. Geol. Sci.* **10**: 21–24.

CHANEY, R. W. 1959. Miocene floras of the Columbia Plateau. *Carn. Inst. Wash. Pub.* **617** (1): 1–134.

COPE, E. D. 1880. A new Hippidium. *Amer. Nat.* **14**: 223.

GAZIN, C. L. 1932. A Miocene mammalian fauna from southeastern Oregon. *Carn. Inst. Wash.* **418**: 37–86.

GREGORY, J. T. 1942. Pliocene vertebrates from Big Spring Canyon, South Dakota. *Univ. Cal. Pub. Dept. Geol. Sci.* **26**: 307–446.

HALL, E. R. 1944. A new genus of American Pliocene badger, with remarks on the relationships of badgers of the northern hemisphere. *Carn. Inst. Wash. Pub.* **551**: 9–23.

MARTIN, H. T. 1928. Two new carnivores from the Pliocene of Kansas. *Jour. Mamm.* **9**: 233–236.

MATTHEW, W. D., and R. A. STIRTON. 1930. Osteology and affinities of Borophagus. *Univ. Cal. Pub. Bull. Dept. Geol. Sci.* **19** (7) : 171–216.

MERRIAM, J. C. 1911. Tertiary mammals beds of Virgin Valley and Thousand Creek in northwestern Nevada. *Univ. Cal. Bull. Dept. Geol.* **6** (11) : 199–304.

—— 1915. New species of the hipparion group from the Pacific Coast and Great Basin provinces of North America. *Univ. Cal. Pub. Bull. Dept. Geol.* **9** : 1–8.

SHOTWELL, J. A. 1955. Review of the Pliocene beaver *Dipoides. Jour. Paleo.* **29** : 129–144.

—— 1956. Hemphillian mammalian assemblage from northeastern Oregon. *Bull. Geo. Soc. Amer.* **67** : 717–738.

—— 1958. Evolution and biogeography of the aplodontid and mylagaulid rodents. *Evolution* **12** (4) : 451–484.

—— 1961. Late Tertiary biogeography of horses in the Northern Great Basin. *Jour. Paleo.* **35** (1) : 203–217.

WILSON, R. W. 1934. A new species of Dipoides from the Pliocene of eastern Oregon. *Carn. Inst. Wash. Cont. Paleo.* **453** (3) : 19–28.

—— 1936. A Pliocene rodent fauna from Smiths Valley, Nevada. *Carn. Inst. Wash. Cont. Paleo.* **473** (2) : 15–34.

—— 1937a. New middle Pliocene rodent and lagomorph faunas from Oregon and California. *Carn. Inst. Wash. Contr. Paleo.* **487** (1) : 1–19.

—— 1937b. Pliocene rodents of western North America. *Carn. Inst. Wash. Cont. Paleo.* **487** (2) : 21–73.